TECHNO-DIPLOMACY

US-Soviet Confrontations in Science and Technology

TECHNO-DIPLOMACY

US-Soviet Confrontations in Science and Technology

Glenn E. Schweitzer

Plenum Press • New York and London

Library of Congress Cataloging in Publication Data

Schweitzer, Glenn E., 1930–
    Techno-diplomacy: US-Soviet confrontations in science and technology / Glenn E.
Schweitzer.
        p.   cm.
    Bibliography: p.
    Includes index.
    ISBN 0-306-43289-7
    1. Science and state—United States. 2. Science and state—Soviet Union. 3.
Technology and state—United States. 4. Technology and state—Soviet Union. 5.
Science—International cooperation. I. Title.
Q127.U6S288   1989   89-8787
338.97306—dc20   CIP

© 1989 Glenn E. Schweitzer
Plenum Press is a Division of Plenum Publishing Corporation
233 Spring Street, New York, N.Y. 10013

Printed in the United States of America

Preface

Techno-diplomacy \tek-nō-də-'plo-mə-sē\ n. (1989)
1: the art and practice of conducting negotiations
between countries with conflicting technological
interests, 2: skill in handling scientific affairs
without arousing hostility, 3: ability to resolve
issues on the frontiers of science and technology
in the direction of peace and not war : TACT[1]

Should the United States help Mikhail Gorbachev succeed? The stakes are enormous. Our human values and our security are on the line.

The USSR and the United States control most of the world's military power, adversarial power that is the product of science and technology. At the same time, our two countries have many common interests in using science and technology both to help raise living standards and to protect health and ecological resources on a global basis. Can the leaders of the two superpowers devise practical approaches for channeling more of the scientific energies of their countries into human-itarian endeavors and away from military pursuits? This is the challenge which must be met by policies of the two countries, policies which meet on the frontiers of science and technology. This is the challenge of techno-diplomacy which can help to ensure that such policies steer our nations away from confrontation toward cooperation, for continued confrontation could lead to war, while cooperation offers the best hope for peace.

The radical changes underway in the Soviet Union are driven to a large extent by the determination of the Soviet leadership to harness science and technology and thereby modernize a sputtering economy. The rewriting of history, the whittling away of the bureaucracy, the promotion of economic entrepreneurship, and the turning loose of artists and writers are all part of an unprecedented attempt to dramatically transform an entire society and to prevent the country from slipping further behind the technological and economic achievements of other nations. Resources are scarce and incentives to reduce military spending are great; the time is right for the two superpowers to reach accommodations that can redirect priorities and change the world.

The United States and the USSR first established diplomatic relations in 1933. On that occasion President Franklin Roosevelt emphasized ". . . our nations henceforth may cooperate for their mutual benefit and for the preservation of peace of the world."[2] Now, 56 years later, our two countries have a new opportunity to transform this hope of the past into the reality of the future; science and technology will be a pivotal factor in how our relationship develops.

*   *   *

The shortage of computers and computer illiteracy at all levels of society prevent the USSR from entering the information age. The Chernobyl disaster and the crumbling of buildings during the earthquake in Armenia have shattered many of the mythologies about the prowess of Soviet engineering. Continued crop failures reflect the limitations of ideologically based agricultural policies.

One-quarter of the world's scientists live in the USSR, but the productivity of most Soviet laboratory scientists is less than 30 percent of the productivity of their American counterparts. Few Soviets have mastered the keyboard for typewriters or computers, and few have access to copy machines. Though a small number work in sparkling new laboratories, most struggle in dreary, rundown facilities which seem to be always under repair.[3]

An astonishing half of the world's engineers work in the USSR. Most are technologically underemployed, however, toiling as technicians and denied the tools or opportunities to apply modern engineering skills. Meanwhile, many automated production lines rely on workers to take over when the control systems malfunction, with the quality of Soviet products unpredictable at best.

For years Western visitors to Moscow and the provinces have likened the lack of amenities in the USSR and the chronic shortage of consumer items throughout the country to life in developing countries. Comparisons of the conditions in the USSR with survival in the Third World have bruised the pride, riled the anger, and awakened the consciousness of Soviet leaders.

Despite economic stagnation of the country, Soviet military might looms as large as ever. The Soviet nuclear arsenal has doubled in the past 15 years. The nation's conventional forces span many thousands of miles. And its spectacular space achievements remain unmatched.

This distorted situation has provided much of the impetus for Gorbachev's plans to reconstruct Soviet society—perestroika. To many Soviets, perestroika seems to be the only hope for reversing the continuing decline of the Soviet Union as a modern nation. Even the Soviet critics of perestroika offer no alternative course of action. Gorbachev is pinning many of his hopes, and indeed his survival, on the effective use of the huge pool of technical manpower—including the managers of military programs—to revitalize the entire Soviet economy from top to bottom.

Meanwhile, Gorbachev dazzles the world with his intellect, charisma, and boldness. Under his leadership the Soviet Union is changing in ways considered unbelievable several years ago.

*   *   *

This book is about science and technology—a central theme of perestroika. It is about Gorbachev and Soviet scientists and about the charting of a nation's course into unknown seas.

This book is also about American scientists and engineers, about our universities, private companies, and government agencies. It is about those Americans who have worked long and hard to blunt the edge of the US-USSR military confrontation by promoting cooperation in programs of considerable importance to both nations and to the world. Frequently these Americans have experienced the lavishness of Kremlin receptions and the excitement of the Bolshoi Ballet in Moscow. But in Washington their importance as a rudder of political stability in the rough turbulence of US-Soviet relations has yet to be adequately recognized.

Most importantly, this book is about power. Science and technology can be equated with power—military power, economic power, environmental power, and the power of new ideas and hope.

The concept of supremacy through military power lost much of its earlier significance shortly after the world entered the nuclear age. Long ago both superpowers crossed the threshold of nuclear capacity to destroy civilizations, rendering additional weapons of little military value. A formidable nuclear deterrent is at the ready in Central Europe, and additional conventional forces in the region appeal only to theorists who fantasize that a Third World War might not involve nuclear weapons. Clearly, attempts to achieve military superiority should become less of a driving force in the development of superpower relations.

Economic power is rapidly dominating the international scene. The influx of high technology products from Asia into all parts of the world shows that economic power is not necessarily derived from military power. International financial dealings around the globe transfer the equivalent of more than $1 trillion across international borders daily; that is where the power which will shape the next century lies.

The Soviet Union cannot remain isolated from international economic developments. The country painfully feels the economic impact of fluctuations in the international price of its oil exports. The nonconvertibility of the ruble embarrasses Soviet travellers throughout the Western world. Also, the USSR is in a quandary as several of its East European satellites sink deeper into international debt.

The indirect effects of the changing international economic scene on the superpower relationship, and indeed on international relationships throughout the world, are profound. Shaping these effects is the revolution in microelectronics; those countries that can obtain and use information on a global scale will be the true masters of their economic future. As former Secretary of State George Shultz frequently observed, the information standard—reflected in the power and compatibility of computer systems—has replaced the gold standard as the basis for international commerce.[4]

National security, or perhaps more correctly national survival, for both countries is also dependent on power to conserve natural resources, to protect the water and air from pollution, to prevent the spread of infectious diseases, and to maintain a healthy population. Such power is fed by the creativity of scientists and engineers. Americans believe that our approach to governance offers the greatest promise for stimulating creativity—an approach which springs from a tradition of free thinking. Now all nations must not only stimulate scientific creativity but must ensure that it is used to promote the interests of all humanity.

*   *   *

Discussions of technological competition between East and West would not be complete without commentaries on the activities of the foreign intelligence services and the internal security agencies of the United States and the USSR—the CIA, the FBI, the KGB, and others. These organizations have many roles, which include keeping track of technological developments abroad and protecting technologies at home. The publicity attendant to clandestine intelligence activities around the beltways of Washington and Moscow has been so extensive that many American travellers to the USSR and Soviet visitors to the United States are sure they are under the ever-watchful eye of an adversary. Needless to say, such activities of the intelligence agencies greatly blur the propriety of many cooperative activities between the two countries and confuse and confound Soviet and American participants alike.

The role of some diplomats and of foreign agents in cat-and-mouse intelligence games should not be confused with the role of scientists in exchange programs. While I was serving as a diplomat in Moscow, our apartment was bombarded with still unexplained microwaves for three years, and a microphone was removed from the wall behind my Embassy desk. Also, I was physically hassled by the KGB on the streets of Moscow, and my family was harassed on a holiday excursion outside Moscow; these personal indignities were Soviet retaliations for FBI encounters with Soviet diplomats in Washington who were apparently engaged in much more sinister activities than I.

In sharp contrast, as a representative of the National Academy of Sciences, which is a nongovernmental organization, I have been treated with great respect in both countries. I have never sensed an effort to distort my position as a promoter of cooperation in areas of considerable interest to both countries. Our scientific exchange visitors only rarely have difficulties; yet once in a while Americans have encounters with the KGB, and Soviet exchange scientists arouse the concerns of the FBI.

*   *   *

Much of the writing draws on my experiences as the first Science Officer at the American Embassy in Moscow in the mid-1960s and on impressions during the past four years as a manager of a large nongovernmental scientific exchange program between the United States and the USSR. During the intervening 18 years I occasionally visited the USSR. More importantly, I was deeply immersed in directing scientific endeavors in the United States during this period. This experience in American research laboratories sensitized me to the sharp differences faced by American scientists working in the United States and their Soviet counterparts conducting research in the USSR.

I have included many personal anecdotes in the text to add a human dimension to the discussions. Hopefully, these personal observations will be received in the spirit which prompted their inclusion. I am just one of many Americans who have tried to stay in touch with developments in the USSR over the years. However, I have been fortunate to have witnessed many unusual events in the United States and in the USSR, and this book offers an opportunity to share some of the most interesting experiences with a broader audience. Given my heavy reliance on personal recollections as the basis for many of my impressions, an academically oriented book would not be appropriate. Thus, footnotes are limited, assertions are manyfold, and conclusions are frequently controversial.

In short, the book is an unvarnished recording of my impressions of developments in US-USSR scientific relations up to the end of the Reagan administration. I have benefited enormously from discussions with the leaders of the National Academy of Sciences, from debates with my associates at the Academy, from interchanges with American scientists and with specialists in Soviet affairs, and from interactions with Soviet officials and scientists over the years. I am profoundly grateful for the insights which these colleagues have provided. Still, the views in this book are my own, particularly those views which may run counter to conventional wisdom about developments in the USSR or which conflict with policies adopted in Washington.

I am very appreciative of the support provided by the staff of Plenum Publishing Corporation in the preparation of the manuscript. In particular, Linda Regan's insights in separating the wheat from the chaff and her numerous contributions to improving the logic and readability of the presentation were invaluable.

Finally, I would like to express my appreciation for the support of my family as I wandered the globe over a period of many years. From our first night in the USSR at the Hotel Bug in Brest on the Polish border in 1963 until the final reading of this manuscript, Janet, Carol, and Diane have been a source of inspiration that has seen me through many cold nights in Archangel, Novosibirsk, Leningrad, and Moscow and many lonely days at the word processor.

Contents

CHAPTER 1

Perestroika, Technology, and the Uncertain Soviet-American Relationship

Mikhail Gorbachev is my friend.
Ronald Reagan

Kremlinologists must retool. With the onset of glasnost, the information drought as to Soviet concerns and intentions has given way to uncontrolled floods of information about deliberations and debates at all levels of Soviet society. No longer is an interview with a "well placed Soviet source" or with a Soviet political dissident the primary objective of Western journalists and diplomats. So many sources and dissidents are speaking out on all fronts that Western observers now have the luxury of emphasizing those Soviet statements which best support their personal perceptions of trends in Moscow and throughout the country.

The recent changes in the Soviet media are startling. Soviet journalists have begun to seize the initiative from the Western press in exposing corruption, inefficiencies, and other weaknesses in the Soviet system. In recent months I have seen new Soviet films and read articles in leading Soviet magazines based on real-life experiences in the USSR which depict prostitution, extortion, and murder with a violence that deserves R ratings.[1] The Soviet Government arranges formal press conferences for Sakharov and other controversial personalities who apparently speak without inhibitions. Name-calling on television among the party leadership has been in fashion since the 1988 party conference.

Leading Soviet officials make so many conflicting statements that confusion reigns at home and abroad in sorting out the party line, if indeed there is a party line. While Gorbachev's statements clearly carry far more weight than the pronouncements of others, even he must frequently modify his views lest he find himself isolated within a party of many conservative forces.

Meanwhile, the spread of activities of foreign visitors in the Soviet Union during the past several years has been remarkable. Western business representatives are seemingly everywhere trying to develop commercial schemes for tapping into the 300 billion rubles of personal savings of Soviet citizens throughout the country. American military inspection teams travel to formerly secret sites. Hotels capable of accommodating foreigners are now appearing in some areas which had previously been off-limits to visitors. Meetings of visitors with refuseniks have become routine. Soviet media coverage of statements by Western leaders and ordinary citizens from the West is extensive and far less selective than in the past.[2]

*   *   *

A good roadmap for positioning recent developments in the USSR within the Soviet commitment to communism is Gorbachev's book *Perestroika*, published in the English language in the United States in 1987 and published shortly thereafter in the

USSR in Russian. Keeping in mind the importance of the American audience, he maintains his vision of the future directions for Soviet society as a natural extension of Lenin's principles. His assessment of problems and opportunities within the USSR as well as on the international scene has provided the basis for many of the policies which he has advocated at home and abroad and has frequently put into practice during the past several years. While glasnost has probably shaken more skeletons in the closet than even Gorbachev anticipated and has prompted additional proposals for reform, Gorbachev has generally followed the perceptions he set forth in 1987.[3]

Complementing *Perestroika* is another book published in 1988 for consumption in the West, *The Economic Challenge of Perestroika*. This book is authored by one of Gorbachev's economic advisors, Abel Aganbegyan. It chronicles many mistakes within specific Soviet enterprises during recent years and forcefully makes the case for the economic reforms which should be introduced.[4]

American economists are quick to point out some of the shortcomings in the rigor of Aganbegyan's economic analyses—at least as we understand economics in the West. Soviet critics complain that his proposals to change organizational responsibilities are simplistic. Aganbegyan admitted to a small group of us in Moscow that he has great difficulty being a "partial virgin" trying to retain the purity of Socialist equality while advocating introduction of capitalist entrepreneurship. Nevertheless, the book sets forth the basic concepts for moving from rigid central planning to self-financing of industrial activities at the enterprise level. Inherent in this approach, argues Aganbegyan, must be a devolution of decision authority to the factory managers and to the workers themselves.

Gorbachev's three-pronged program of glasnost, perestroika, and democratization has received enormous publicity within the USSR as well as notoriety within Western intellectual circles. All branches of the Soviet media rightfully boast of their new openness. Many bureaucrats have been prematurely retired. As the concepts of multiple candidates for elected positions and of limited tenure for senior officials are being introduced within the government, those most directly affected by the democratization process are the party faithful, who in the past have benefited from the favoritism of a political spoils system. Also, the traditional meddling of the party in the micromanagement of many social and economic activities is being eroded as economic factors receive greater weight. Meanwhile, most Soviet citizens have yet to be personally touched by the program of reforms.

In 1988 a new government research institute for public polling was established in Moscow. An early effort assessed public reaction to glasnost. This poll found that few Soviet citizens can define glasnost, and even fewer can give examples of glasnost in action. Most prefer to "read, listen, and watch" rather than "personally and directly participate in" the process of democratization. According to the poll, glasnost has been most popular with the well educated and with residents of the two largest and most sophisticated cities, Moscow and Leningrad.[5]

*   *   *

For the first time the Soviet leadership seems to appreciate the depth of the systemic problems that hold back the Soviet economy and to recognize the failure of their science and technology in areas outside military systems and space exploration. Soviet leaders are aware of how commonplace high technologies have become in many countries while recognizing the minimal Soviet participation in the international process of technology diffusion. They increasingly appreciate the limits of military power—in the superpower relationship and in addressing problems in the Third World, and they see great benefits from arms control agreements that can lead to reduced spending on armaments. They have truly been awakened by the German youth who in 1986 penetrated the supposedly sophisticated Soviet air defenses all the way to Red Square in a light airplane, by the consequences of their isolation from the information revolution, and by the eight billion ruble price tag for the cleanup of Chernobyl.

Meanwhile, the Soviet population reacts cautiously to the rewriting of Soviet history and to the growing discrepancies between Marxist theory and Soviet practice. But then Marx was a foreigner, they note. They are frequently confused, occasionally amused, and sometimes bored by the strong drums of self-criticism. In general, however, Soviet attitudes seem receptive toward the changes being introduced into society from the top. Soviets are proud that Gorbachev is receiving international recognition as a world leader—as *Time*'s man of the year (1987), as an equal with the president of the United States and the prime minister of Great Britain, and as an advocate of policies aimed at global harmony.

The Soviet population sees bright sides of perestroika. They welcome the return of Soviet soldiers from Afghanistan. They are glad that corruption is being exposed. They are pleased that the laggards are to be put in their places. The Soviet population knows no alternative to communism, and they have backed Gorbachev's efforts to improve the system. At the same time, the workers are restless. Shortages of consumer goods are severe, and the promised tools to improve productivity are nowhere in sight. Frustrations will undoubtedly continue to burst forth in street demonstrations from time to time as Gorbachev struggles to quell the population's impatience with delays in realizing the economic promises of perestroika.

The concept of entrepreneurship, which is a key to perestroika, will require many years to catch on in a significant way. Managers of large state enterprises are not accustomed to having responsibility for turning a profit, let alone being motivated to organize the efforts of the workers solely for profit. Presumably with the help of the central government, enterprise managers will begin to mobilize the necessary talent which can cut through the system and ensure reliable sources of raw materials at reasonable costs. They will also have to find customers willing to pay fair prices, and they will have to deliver products of adequate quality in order to prevent the loss of markets to competitors.

These economic concerns so familiar in the West are only a small part of the problem of entrepreneurship in the USSR. Enterprise managers will have great difficulty in refusing to comply with directives from Moscow or from local authorities for emergencies, for exceptions, and for political favors, though compliance may hurt their profits. Unproductive workers cannot be fired if they have nowhere else to find employment. Managers will be reluctant to reduce the many social services traditionally provided to the workers—housing, transportation, schools, camps, medical facilities, cultural events, political education; but these services have little economic return for the enterprises.

Are these types of supplementary activities to be subsidized by the state? What alternatives exist if domestic cost accounting is to be brought into line with international cost accounting? Such alignment seems essential if the Soviets are to be players in world trade.[6]

*   *   *

Parallel to the internal reforms have been Gorbachev's shifts on international security concepts. He has linked national security with mutual security, pointing out that a nation's own security cannot be ensured without taking into account the security of other nations. If other states feel insecure, they will take military steps to redress the balance, steps that in turn provoke further military responses by their potential adversaries.

Gorbachev has clearly articulated the need to broaden the traditional Soviet concept of national security from sole reliance on military strength to include both political and economic strength. He has recognized the importance of a genuine interdependence among nations in international politics and economics. He is apparently moving away from the earlier Soviet concept of international division of labor which called for each nation to rely primarily on its particular economic strengths; this philosophy had provided a basis for arguing against discriminatory trade practices of the West. Finally, he seems to be downplaying the importance of national liberation struggles in the Third World, due at least in part to the economic drain on the USSR from supporting such struggles.

Skeptics in the West give little credence to Soviet rhetoric concerning new approaches to national security as long as the USSR continues to modernize its weaponry and diffuse arms around the world. They are impatient. They welcomed recent steps toward limited cuts in military expenditures by the Soviets and reductions in the size of their army, but they want action and not just promises. These voices in the West argue that the Soviets will only feel secure when the West does not have the capability to resist Soviet aggressive ambitions.[7]

*   *   *

As to better use of science and technology, Gorbachev's program of reforms has two thrusts. First, many obsolete Soviet technologies must be brought up to the world standard of the 1960s and 1970s, and eventually to a contemporary standard, in industries ranging from automobile manufacturing to textiles. Second, many new technologies in the fields of electronics and advanced materials must be mastered if the USSR is to enter the information age that is spreading rapidly throughout the world.

Gorbachev vividly recounts the failures of past Soviet policies of heavy reliance on importing technologies from the West. This dependence on Western capabilities reduced incentives in the USSR for the development of Soviet technologies. Then the policies became entwined in embargoes and bans imposed by the West. While still calling for the use of foreign technologies, he now advocates accelerated indigenous technological development which can be meshed with the best technologies available from abroad.[8]

The USSR is not alone in seeking technology from all available sources. Countries throughout the world are pinning their hopes for a better standard of living and for protection of their national boundaries on their mastery of science and technology. In the United States we consider new technologies as the key to ensuring international competitiveness for our manufactured products and as the cornerstone to maintaining military superiority in many regions of the world. As previously noted, improved science and technology in the USSR are to be at the forward edge of perestroika while also strengthening capabilities to withstand new military challenges from the West. Meanwhile, the developing countries perceive science and technology as the only bridge across the ever-widening rivers of poverty; these waters become more treacherous with the continuing population explosion, the ravaging of the tropical forests and other natural resources, and the emergence of AIDS as a threat in many countries.

The rapid growth of computer technologies excites leaders and opens new doors to political and economic power. International capabilities to develop and introduce the latest electronic devices are rapidly spreading. Industrial firms in the United States, Western Europe, and Japan have high-tech facilities around the globe; and local entrepreneurs are also emerging in Taiwan, Hong Kong, Korea, and Latin America with capabilities to produce electronic clones. The research and development networks and the manufacturing complexes of many multinational companies operating in electronics and other high-technology fields simply no longer respect national boundaries, despite disparities in regulations controlling financial, production, and export activities in different countries. In general, Western technologies belong to companies and not to countries. These companies are now dispersed as never before.[9]

The pressures for commercialization today of scientific discoveries of yesterday have diminished the distinctions between research, development, and application. While rapid introduction of the results of research into practice has been an oft-repeated axiom for many years in almost all countries, only recently has the economic importance of reducing the time lags in the research and development cycle really hit home. Several centuries elapsed between Newton's discoveries of the laws of gravitation and the first artificial satellite. Now, the development cycles for new generations of computers have been compressed to periods measured in months. The cycles for new generations of software are sometimes measured in weeks. Even in agriculture, where plant breeders have usually thought in terms of decades for developing new varieties, biotechnology promises to truncate the lead times for new products of commercial value.

Scientists engaged in basic research have always been worried that the quest for quick payoff from science will reduce support for their activities. They contend that only better understanding of fundamental processes will result in major break-throughs. Such understanding is obtained through research which rests on unprogrammed trial and error to gain new knowledge that in the short run may be totally divorced from the development of new products. The most important aspect in the application of a new technology is a good theory, they argue. They do have a point. However, in the scramble for limited research funds, tensions will persist among the basic researchers and those forces which seek more immediate payoffs from science.

Meanwhile, military establishments in all countries have increased their interest in how science and technology can best serve their needs. In the United States, the Department of Defense now funds over 60 percent of all research and development supported by the US Government or 30 percent of all research and development conducted in the United States. This means that the Department of Defense directly and indirectly employs a very large portion of our science and technology workforce. The areas of interest to the Department of Defense span all scientific disciplines from anthropology to zoology. Yes, anthropology provides background on the likely behavior patterns of enemy forces, and zoology gives insights as to possible battlefield conditions in remote areas.

With each passing day the distinction between science and technology for nonmilitary uses and for military uses becomes less clear. Following World War II, many civilian technologies were spin-offs from military activities—jet engines, communication systems, nuclear power. Now the reverse is also true, with many modern military systems being the beneficiaries of Western civilian research and development efforts—computers for controlling battlefield systems, plastics used in aircraft and missiles, and satellite technology. In the future, military establishments throughout the world will probably be interested in every aspect of science and technology. Of course, the military interest is most intense with regard to those technologies which can be used in weapons systems—radars, heat-resistant materials, undersea acoustic devices. Such technologies which have both military and civilian uses are called dual-use technologies. It is in the area of dual-use technologies that most of the policy debates concerning East-West trade and technology relations generate sparks. To what extent should the United States forgo opportunities to sell its products abroad in order to prevent dual-use technologies from falling into Soviet hands? Given the availability around the globe of dual-use technologies, can these technologies indeed be protected?

Also of particular relevance to East-West cooperation is the rapid emergence of big science projects. Big science usually means very expensive science. It attracts top scientists and involves very sophisticated instrumentation: spacecraft, particle accelerators, astronomy observatories, and undersea research vessels. Clearly no single scientist has the capability or the financial resources to grapple with big science, and large facilities and teams of scientists supported by governments are essential. Now the calls for international sharing of the costs and pooling of technical talent in big science are at an all-time high. Potential Soviet participation in such international arrangements is always a controversial topic. What will they bring to the program, and what will they take away?

A new dimension of big science is developing rapidly, namely, the scientific aspects of changes in the very nature of the biosphere brought about by pollution and destruction of natural resources. Ozone depletion, the greenhouse effect, and acid rain are now household words. The interest in international cooperation in addressing these environmental problems is clear, and many research programs are being initiated, nationally and internationally. The humanitarian nature of such efforts should be of interest to everyone on earth. Yet concern over the desirability of sharing technologies even for humanitarian purposes looms in the background. Military analysts are wary that such sharing could erode our technological lead in both the economic and military spheres. For example, will use of satellites to scan pollution on a worldwide scale compromise photographic technologies which detect missile sites? Will development of mathematical models of complicated environmental processes require international sharing of advanced computer technologies?

*   *   *

While developments unfold in the USSR with unaccustomed speed, the process for formulating and redirecting US foreign policies remains complicated and slow, particularly with regard to policies affecting the relationship between the superpowers. Many highly influential Americans and institutions of all political persuasions are directly involved in shaping new US policies. Many other countries are affected by the superpower relationship and understandably attempt to influence how that relationship develops. Also of great importance, most Americans believe that they have a very personal stake in the character of the relationship between the two countries which could easily affect the future of humanity. They want to be certain that changes will indeed improve prospects for survival. Thus, it is not surprising that we are witnessing dramatic changes in the USSR which are clearly moving at a much faster rate than adjustments of US foreign policy to these changes.

In late 1987 former Secretary of State Shultz, speaking to an audience of American foreign policy specialists, articulated a sophisticated assessment of global developments surrounding the superpower relationship. The increasing economic dependencies among countries, particularly those involving the Asian countries, together with the tremendous strides in the speed of diffusion of information are encouraging democracy, openness, and freedom, both within and among nations. He argued that those nations which have adopted greater freedom in their marketplaces and in their political institutions are the ones which are advancing. Those that suppress innovation and information are falling further behind. In Shultz's view, Gorbachev's "new thinking" is an effort to compensate for historical Soviet weaknesses in these areas of critical importance as the Soviets lose influence around the globe.[10]

As to whether the US Government hopes Gorbachev will succeed or fail, Shultz and other administration spokespersons have been hesitant to take a position. They invoke diplomatic niceties of not interfering in the internal affairs of the USSR. However, they note that the USSR is clearly seeking a breathing period of 10 to 20 years devoid of military threats from abroad in order to be able to concentrate on improving its domestic economic situation.

Many Western skeptics contend that while the changes within the USSR are dramatic on the surface, they are little more than cosmetic in reality. Food lines remain long, consumer goods remain in short supply, agricultural productivity has not improved, and the industrial base continues to age. They emphasize that the KGB is still out of control. These skeptics perceive the proposed changes on the domestic front as unworkable and easily reversible. They still see an aggressive Soviet posture in the Third World, notwithstanding the pullout from Afghanistan, and Soviet military and economic domination of Eastern Europe. They argue that if significant changes are to occur, they will be primarily the result of military and economic pressure applied by the United States. This pressure, they conclude, must be continued to eventually break the Soviet system.

However, to many other Americans, Gorbachev's ascendancy has brought a great deal of hope that life will be different in the USSR and that the superpower relationship therefore can also be different. They identify very easily with the concepts of glasnost, democratization, entrepreneurship, and human dignity being espoused by the Soviet leadership. These are Western concepts, and it would appear that we are winning the war of ideologies. Soviet support of inspection for arms control agreements and advocacy of deep cuts in nuclear arsenals also hit responsive chords.

*   *   *

Early in the Reagan administration, the United States proposed as a negotiating tactic dividing relations with the USSR into four areas, namely, arms control, human rights, regional problems, and bilateral issues. The Soviets eventually agreed to the structuring of discussions into these areas. However, disagreements arose over the relative importance of the areas and the interrelationships among them.[11]

The impact of the four Reagan-Gorbachev summits on the official attitudes of the two countries was substantial. Both sides agreed on the importance of continuing bilateral dialogues on a wide range of issues in the nuclear era. To quote Gorbachev, "There is no getting away from each other." They recognized the difficulty in managing the superpower relationship in view of contrasting philosophies, political systems, and national interests. Candor dominated discussions. Both sides appreciated the rapid political, economic, and technological changes throughout the world that affected both countries; and they indicated a willingness to entertain new ideas and approaches.

Clearly, both sides considered the area of arms control of paramount importance. Enormous effort was expended during the four summits and at many other forums to try to come to closure on a number of arms control issues. (At the summit in Geneva, the US negotiating team occupied 127 rooms at the Intercontinental Hotel.) Agreement on the INF treaty and substantial progress toward improved verification measures that could lead to ratification of two treaties restricting underground nuclear testing highlighted the arms control progress during the Reagan administration. The international debate on many other military issues also reached a new level of seriousness. And science and technology were on center stage during the discussions of SDI and of the verification aspects of all types of agreements.

Reagan's public view of the "evil empire" changed, and Gorbachev acknowledged that maybe the empire had been evil in the past. Both leaders recognized the growing impact of scientific progress on a broad range of issues confronting the two countries.

Still, human rights has remained of special concern to the United States. Much of our attention has been directed to the plight of Soviet Jews, and particularly their imprisonment for political reasons and denial of their emigration from the USSR. Many of the Jewish dissidents and refuseniks have been scientists, and the reaction within the American scientific community to Soviet infringements of human rights has been substantial. Concern has also been aroused in the United States over nationalistic demonstrations in the Baltic republics of Estonia, Latvia, and Lithuania—areas yet to be legally recognized by the United States as being a part of the USSR. In addition, as glasnost spread into the provinces during 1987 and 1988, disturbances in Armenia, Georgia, and other republics made headlines in the Western press.

Discussions between the two governments of regional political problems have covered the globe, with special emphasis on southern Africa, Central America, and the Middle East. The United States argues that the USSR has little business in these areas, particularly in view of its emphasis on military involvement. The USSR contends that, as a world power, it has as much right to participate in the activities of these areas as does the United States, whose military aid programs reach many corners of the world.

Discussions about cooperation in science and technology take place in the forums directed to bilateral issues. Negotiation of intergovernmental agreements to further some aspects of scientific cooperation has always been an item on the agenda during preparations for summits. At the Geneva summit in 1985, cooperation in fusion research was in the spotlight. In Washington in 1987 the leaders underscored their interest in studying the impacts of man-induced environmental changes on the climate. In Moscow in 1988 they highlighted cooperation in the exploration of Mars.

A number of official scientific exchange agreements were terminated during the early 1980s due to US disapproval of Soviet actions in Afghanistan, Poland, and elsewhere. However, by 1988 the Reagan administration had initiated or restored more than 10 intergovernmental agreements involving science and technology cooperation and had seriously discussed other areas for future cooperation.[12]

A particularly contentious issue within the US Government during the late 1980s was a proposed intergovernmental agreement for cooperation in basic scientific research—research considered far removed from industrial applications. Several agencies argued for three years against such an agreement, which they contended would legitimatize Soviet access to American scientists who could be exploited by Soviet intelligence forces. Finally, after extensive debates within the US Government, American and Soviet negotiators developed a text of the agreement prior to the summit in Moscow in June 1988. Still, on the eve of President Reagan's departure for Moscow, conservative forces within the White House objected, and the agreement was shelved. Six months later the agreement was revived and finally signed in early 1989.

Bilateral economic relations also command considerable priority during discussions between the two governments at summit meetings and at many other diplomatic sessions. The current level of trade between the two countries is very small. In comparison with American trade with many other countries, it is almost negligible. Nevertheless, American business executives continue to participate in both official and unofficial discussions with Soviet organizations, confident that eventually there will be opportunities to take advantage of the large natural resource base in the USSR and to penetrate the Soviet consumer and industrial markets in a more serious manner. The new Pepsi Cola commercials on Soviet television add considerable spice to conversations among American business representatives grasping for indications that the Soviet market is opening up. The Soviets have been seized with mastering Western approaches to marketing, which they believe will help internal economic growth within perestroika and aid their linking to the global market. American marketing consultants are popular visitors in Moscow.

As expected, the USSR is very interested in trade arrangements which provide access to the technological achievements of American industry. Ideologically, the Soviets continue to argue that Western economic discrimination against the USSR through embargoes and high tariffs inhibits the prospects for peace. Meanwhile, they chip away at trade barriers by entering into as many favorable commercial deals with Western firms as possible, and they are exploring how they can become full-fledged members of international trade and financial organizations such as the General Agreement on Tariffs and Trade (GATT) and the International Monetary Fund (IMF).

* * *

As we have seen, perestroika, the development and protection of advanced technologies, and Soviet-American relations on many fronts are inextricably linked. The Bush administration must address these issues collectively with policies which advance the recent positive trends in the development of the bilateral relationship.

The changes in the USSR are dramatic. Some cannot be reversed, since many genies are out of many bottles. At the same time, fundamental reform of the Soviet system will not come easily. The year 2017, the centennial of the revolution which led to the Soviet state, seems a realistic target date for the emergence of a new type of Soviet state.

Changes in the upper echelons of the Soviet leadership are now occurring with regularity. While some long-term political leaders remain, new faces increasingly dominate political photo opportunities in the Kremlin. These changes seem to reflect a determination to prepare for the long haul. The many Western predictions of several years ago that Gorbachev's tenure should be measured in months have lost credibility as his personal allies reach the higher pinnacles of political power and as he continues to receive broad international acclaim.

The safest course is to predict that perestroika will fall on hard times. Given, however, that the inflexible Soviet system has managed to withstand both the shocks of war and dislocations during peacetime, it may well adjust to this latest call for change. Indeed, the Soviet leadership may find that the population adjusts too well to newly found freedoms.

In contrast, the American approach to domestic and foreign policies is generally perceived to be quite flexible but is actually rather rigid when discussions turn to our adversarial relationship with the USSR. A significant factor is the hundreds of billions of dollars of defense contracts which are at stake should there be changes in our defense posture. And our defense posture is at the heart of the Soviet-American relationship.

During the 1990s the issue of protecting American technology unquestionably will be a key consideration in the policy debates over our relations with the USSR. This issue is central to our trade relationship. It constrains many of our scientific exchange programs. And it certainly stirs emotions throughout Washington.

The prevailing bureaucratic mind-set has been, "We have the technology. They want it. Don't give them anything." The minority voices have argued, "We aren't the only ones who have it. It's better if they buy it from us than help the foreign trade balance of other Western countries. Let's not control the uncontrollable, and let's relax controls when we can obtain something in return." The debate is not whether we should control militarily relevant technology. We should. The debate focuses on which technologies should be controlled and how they should be controlled.

The superpower relationship is critical to global survival. How can we best respond to the advent of future technologies which will drive this relationship? The chapters which follow explore many dimensions of scientific and technological development to help answer that question. We will see that the only sensible choice involves walking a tightrope of cooperation with the Soviets on the high-tech frontier—walking the tightrope of techno-diplomacy.

CHAPTER 2

Changing Concepts of Military Power and National Security

Our policy is one of deterrence.
Former US Defense Secretary Frank Carlucci

Nuclear deterrence has long outlived itself.
Soviet Defense Minister Dmitri Yazov

Since World War II the wrestling for military supremacy has played center stage in Soviet-American relations. This competition dominates debates in Moscow and Washington over the allocation of national resources and molds the foreign policies of the two countries toward allies in Eastern and Western Europe. Assistance and coercion in the Third World are often designed to prevent military expansion by the other superpower.

The notion that this competition can be replaced by East-West cooperative endeavors in the near term is wishful thinking. However, continued progress in reaching agreements in arms control and in reducing military involvement of the superpowers in Afghanistan, Central America, and other lands beyond their borders could dramatically change the character of the competition. Most importantly, the preoccupation of both nations with the use of military power to shape the bilateral relationship between the two countries can and must diminish.

The impact of the electronics revolution is forcing the USSR to face up to its backwardness in many military and civilian areas. The Soviet leaders see no alternative but to enter the international economic arena and become a significant participant in this technological revolution. They finally realize that to be acceptable internationally, the USSR must become less hostile in words and deeds. The rapid diffusion of electronic and related technologies throughout the world is also having a profound economic effect on the United States. "International competitiveness" has changed from a slogan to a matter of necessity.

The United States needs to move away from its preoccupation with the East-West military axis as the principal determinant in its foreign policy. Our nation should recognize more fully the latent impacts of technological advances in Japan and Western Europe on our economic interests at home and abroad, and we need to devote more energy to reducing the continuing turmoil in Latin America. A more constructive relationship with the USSR could permit greater attention by the United States to these and other pressing international issues.

Cooperation between the United States and the USSR in science and technology can help to build trust between important specialists of the two societies who understand many of the relationships between militaristic and peaceful pursuits. Politically, the time is right for such cooperation to expand in scope. As we shall later see, mutual scientific benefits from cooperation are already on the rise, and such tangible results of cooperation can help shape a brighter vision of the global future.

However, within both countries, the vested interests in maintaining the military competition penetrate all aspects of society. The stereotypes of the United States and the USSR as arch enemies bent on destroying each other have become accepted as conventional wisdom. Most Americans are convinced that the Soviets will never back away from their historical commitment to spread communism throughout the world, relying on force when necessary. The Soviet citizenry remembers the ravaging of Soviet soil during World Wars I and II by invaders from the West. They remain very nervous that the Western powers harbor ambitions for still another attack on the Russian homeland. Meanwhile, powerful voices on both sides claim that strong military capabilities built on ever-advancing technologies in general, and nuclear weapons in particular, have preserved world peace for more than 40 years and that any lessening of military preparedness is therefore foolhardy.

The military leaders of both countries play major roles in shaping the East-West relationship. The choices of new weapons systems and the decisions on the deployment of military forces at home and abroad are the most important levers they have with which to influence this relationship. At the same time, far-flung military exercises provoke minor skirmishes between American and Soviet military forces patrolling along the Siberia-Alaska frontier, in the Mediterranean Sea, and in the Indian Ocean. The mock battles between American and Soviet jet fighters depicted in the popular American film *Top Gun* are much too close to reality.

Flexing muscles of military technology at international air shows and during naval port calls buttresses the political interests of both countries. Using military equipment to break international speed, endurance, and other performance records is often a peacetime goal of the generals. In short, the confrontation between the two military giants is alive and thriving.

Technology is the fuel which propels the race for new, improved, and more expensive weapons systems to offset perceived advantages of the adversary. At the same time, technology advances the design and production of more reliable systems and systems which are easier to control; such systems are often touted as less likely to provoke unintended military exchanges. Also, surveillance technology to check on military actions deep inside the other country is opening the doors for arms control agreements which require reliable verification technologies. Such agreements should lessen the need for building up military forces, although some factions in both countries label this argument specious and dangerous.

*   *   *

Gorbachev's "new thinking" directly challenges many existing precepts of the balance of military power. Even skeptics who dismiss his concepts as a giant hoax to lull the West into a state of complacency admit that his intellectual arguments depart sharply from the tired Soviet rhetoric about the superiority of the Soviet state. Before turning to Soviet concepts, a few comments on American thinking will help clarify the evolving context for Soviet pronouncements, actions, and reactions.

For several decades, enlightened leaders of the Congress and the Executive Branch of the United States have worked hard to convince the American public that our national security extends far beyond the size and capabilities of military forces. The timeliness and appropriateness of American political and economic responses to international security crises, they point out, are critical if our interests are to be protected. Still, many Americans nurture the belief that if all else fails in resolving disputes, the United States can simply use or threaten to use its extensive military power to solve any international crisis. This is no longer true, if it ever was. In recent years we have been sufficiently bruised from military interventions in Southeast Asia, the Middle East, and Central America that this lingering conception of the past can now be unequivocally dismissed.

The state of the American economy is an important cornerstone for the entire international economic and financial system. International security conflicts are increasingly entwined with economic problems around the world, and particularly disturbances within the developing countries which can be triggered by changes in the international value of their currencies, losses of international markets, problems of foreign debt, and fluctuations in foreign aid. While the causes of local economic crises cannot be easily traced across international boundaries, it is clear that the health of the American economy plays an important role both in our own national security and in the security of other countries.

Newly espoused Soviet concepts of the major factors affecting national security mirror concepts that are regularly debated in the West. Soviet specialists articulate the interplay among the economies of the world, recognize that military strength does not necessarily equate to political influence, and frame the concept of interdependence among nations in a very broad context. Despite these awakenings, the Soviets join us in having difficulty in fully understanding how military, political, and economic considerations are intertwined in promoting national security.

The Soviet domestic economy is in such dire condition that the USSR desperately needs some relief from external military pressures, which in recent years have exacted increasing levels of financial resources for Soviet military systems. With regard to the international economic scene, their position is also tenuous. They can continue to ship oil, gas, gold, fir, timber, and other raw materials to the international market; but these resources are not without limit, and their value as primary goods has slipped in relation to the value of finished goods on the international markets. At the same time, they have much to gain through more direct involvement in global affairs, particularly in drawing on technological achievements and management know-how of the West. Looking to the future, watchers of the Soviet scene will continue to see a huge unsatisfied consumer market, large quantities of undeveloped mineral resources in inaccessible areas, and a large scientific workforce that needs a stimulus to raise productivity.

As to more immediate military concerns, Gorbachev emphasizes that a nation's security is directly affected by how an adversary judges its own security. If one nation is overwhelmingly strong and perceived accordingly, the adversary will attempt to correct the situation with military buildups which in turn will eventually threaten the security of the first nation. Thus, the logic goes that for the USSR to be militarily secure, the United States must also consider itself militarily secure; neither side should take steps which significantly alter the perception that both nations are secure lest a new phase of the arms race ensue to correct the situation.

Related to this concept of mutual security is Gorbachev's contention that nuclear weapons must be eliminated. He does not accept the notion that some minimum levels of nuclear weapons are needed to deter war. Rather, he argues that as long as nuclear weapons exist, the possibility of escalation of violence to the level of global destruction remains. He apparently accepts the concept of deterrence based on conventional weapons which have both an attack and a counterattack capability. However, this concept becomes clouded when considering how the superpowers, if devoid of nuclear weapons, would counter threats from other countries that independently develop nuclear weapons.[1]

*   *   *

Such total elimination of all nuclear weapons has never entered into serious strategic thinking in the United States. As an extreme case, in 1981 during a visit to the headquarters of the

Strategic Air Command near Omaha, I asked a leading general, "How many missiles do we need to maintain a credible deterrent?" His reply was, "As many as we can build." Several other generals echoed this sentiment, apparently believing that one side could overwhelm the other—politically if not militarily—if that side commanded a significant numerical advantage in the number of deployed strategic weapons. I wondered what had become of the concept of "overkill." Strategies which call for destroying the adversary more than once surely seem wasteful.

Fortunately, few thoughtful Americans share the logic of simply producing as many nuclear weapons as possible; certainly other military officers whom I have known for many years do not have such views. However, as the numbers of nuclear weapons on both sides have multiplied, logic has taken a back seat to the complexities of military "targeting" requirements to cover every conceivable contingency—requirements that call for many weapons indeed. Thus, as one side produces more weapons and introduces defensive countermeasures against nuclear attacks, the other side responds with new targeting requirements, and the buildup of offensive weapons continues.

In the mid-1970s I listened to former Secretary of State Henry Kissinger speak out forcefully against the continued buildup of strategic forces. At that time, he argued that 500 American warheads which could survive a first strike against the United States were more than enough to retaliate in a manner that would immobilize any nation that was foolish enough to launch a first strike. While Mr. Kissinger's views have become less clear in recent years, the premise he expounded 15 years ago, that neither country needed the large arsenals then in their possession, remains valid today as the nuclear arsenal of each side exceeds 20,000 warheads.

For more than 25 years, US strategic forces have rested on the "triad" concept, involving nuclear-tipped missiles on submarines, missiles on land, and nuclear weapons deployed on aircraft. Thus, each of the three military services—the army, navy, and air force—has its own strategic forces, and each of the services also has numerous ideas about how it could use more strategic weaponry of many kinds. The origin of the triad approach can be traced at least in part to interservice rivalries following World War II; the army, navy, and air force each wanted a major role in the development and deployment of nuclear weapons. While great strides have been made by the Pentagon to integrate the interests and capabilities of the three services, these rivalries among the services persist. They give rise to redundancies of quantity, mission, and roles for the nuclear weapons of each service.

The principal rationale for the triad concept is that the United States should not put all its eggs in one basket or even two. If the Soviets develop effective countermeasures against one leg of the triad or if unanticipated technical problems arise in one of the legs, two others remain. However, in strengthening each leg of the triad over the years, our experts have developed a variety of weapons systems including different types of aircraft carrying various types of warheads and different types of missiles with diverse deployment technologies. In effect, there are many reinforcement bars in each leg to help ensure the viability of that leg. Thus, while the triad concept still draws strong political support, it has lost the technological simplicity that made it so attractive in the first place.

Perhaps the triad concept remains the best approach for achieving our strategic objectives; perhaps two heavily reinforced legs would serve us better; or perhaps two reinforced legs and one thin leg would be preferable. As arms control becomes a much more serious dimension of national security, the approach to strategic weaponry needs careful reassessment. The point of departure should be the strategic objectives in a changing world without the constraint of sanctity in the need to preserve a role for each leg of the triad.

Since the 1950s, Soviet rocket forces have been proudly on parade in Moscow and in other Soviet cities at holiday occasions, as the Soviets have repeatedly showcased for the world their military capabilities to reach any adversary anywhere on the globe. Initially, the US Government overestimated their capabilities, both in terms of numbers of missiles and of capabilities of the systems. Indeed, questions arose whether the Soviet rockets on display during the early days were really functioning rockets or whether some of them were mock-ups. While viewing the rockets on the streets of Moscow from a distance of several feet in the 1960s, I was surprised by the many unexplainable cracks and other imperfections in their external appearance, particularly around the bolts, joints, and engine casings. They simply did not look like rockets that would survive on long journeys. In any event, by the 1970s the Soviet capability in rocketry was well established, and our remaining uncertainties related to the details of their designs.

Both sides profess that they do not intend to launch the initial missile or bomber attacks that could trigger a nuclear war and that their strategic weapons will be used only in a retaliatory mode. Neither side believes the other and can cite indicators of latent first-strike intentions. The Soviets point to the US insistence in reserving the right to use nuclear weapons first if absolutely necessary and to the positioning of US aircraft and submarines near their shores. Our government points to past aggressions of the Soviet Union and to Soviet underground shelters where their leaders could retreat to survive a US nuclear response to a Soviet first strike.

Each side argues that it must use the most advanced technologies to develop systems which will enhance East-West political and military "stability" and thereby reduce the risk of war.

However, concepts of stability are often conflicting and confus-
ing. One American concept is that military systems which appear defensive and not provocative to an adversary and which
avoid the possibility that an adversary will mistake defensive
measures for preparations for war should be promoted. But one
person's perception of defensive measures can be another's perception of a threat, and interminable debates ensue over which
systems are stabilizing and which are destabilizing. Debaters
often end their arguments with the conclusion that if the military system belongs to their side, it adds to stability. If it belongs
to the other side, it is destabilizing. For example, during the
1970s, Soviet efforts to develop antiballistic missile capabilities
were criticized in the United States as destabilizing. Yet the
same critics argue that our SDI program is a move toward more
stabilizing systems.

At the 1986 Reagan-Gorbachev summit in Reykjavik, near-
term reductions of 50 percent of the nuclear missile forces of the
United States and the USSR were considered, along with the
objectives of subsequently eliminating all ballistic missiles and
eventually all nuclear weapons. Reportedly, an approaching
consensus on these concepts fell apart when Reagan refused to
compromise on the testing of SDI while Gorbachev insisted that
restraints on SDI testing must be a component of any serious
reductions in strategic weaponry.

Many political and military leaders in the United States and
Western Europe were subsequently relieved that such a far-
reaching agreement was not reached, with or without the inclusion of limitations on SDI. Conceptually, they had difficulty
believing that the elimination of US and Soviet nuclear weaponry would promote stability and keep the peace. They considered
a total restructuring of the military fiber of the leading nations of
the world in a short period to be not only dangerous but politically disruptive and indeed impossible. Probably of greatest sig-

nificance, they feared that the discussions were based more on Reagan's personal convictions than on sound strategic analysis.

Many American commentators strongly criticized Reagan's performance at Reykjavik, primarily on the grounds that he was unprepared to discuss such weighty issues which were unexpectedly placed on the agenda by the Soviets. Despite the presence in Reykjavik of all of the top American advisers in the field of national security, continued the argument, arms control issues are so important that only positions that had been carefully developed through the technical agencies within the United States and through NATO should have been discussed.

As noted earlier, the vested interests in maintaining current armament levels are very strong in both countries. At the same time, pressures to reduce defense expenditures are increasing, and the morality arguments put forth in support of proposals to reduce nuclear weapons attract ever-larger followings. The views of the many participants in the normal processes of government considering far-reaching proposals for arms control will always be in sharp conflict. Thus, at least in the United States, there is no substitute for leadership by presidents who have carefully weighed the many divergent views in breaking out of the bureaucratic conservatism within the government agencies and in setting the framework for meaningful negotiations. Whether such leadership went too far or not far enough at Reykjavik will be debated for decades.

At Reykjavik, Reagan and Gorbachev projected a vision of a world with no nuclear weapons. They apparently wanted to agree on this goal and a timetable for achieving it. This seems unrealistic, but other far-reaching goals which will avoid nuclear annihilation are needed. However, we cannot rely simply on the traditional processes of government or of consultation among allies to generate the initiative for establishing such goals. Our government must consult in Washington and abroad, but we

Americans must look to the president to stimulate visionary thinking about national security.

"If we know where we are going, we have a chance of arriving at our destination. If we don't know where we are going, we have no chance of arriving." This timeworn saying is most appropriate with regard to national security. Eliminating the need for most nuclear weapons by the end of this century is a reasonable destination point in the area of strategic weaponry.

*   *   *

The close relationship between nuclear force levels and the plans of both the United States and the USSR to retain strong nonnuclear military capabilities in Europe has been recognized for many years. However, only now, as reductions in nuclear weapons become a reality, is this relationship receiving the attention it deserves.

Since the end of World War II, Central Europe has been free of serious military conflicts. Of course tensions have sometimes run high, particularly with regard to the status of Berlin. Many incidents related to espionage on both sides of the East-West border and to the escape of refugees from behind the Iron Curtain have also stirred political anxieties. Even today, many of us receive icy stares from well-indoctrinated border guards from the Eastern bloc who do their best to make us feel uncomfortable as we cross the border separating West Germany and Czechoslovakia and as we travel between East and West Berlin. Nevertheless, military forces of the East and West have not engaged in direct combat despite the large numbers of forces deployed in close proximity to the borders. If continuation of the status quo is an indicator of stability, then the border between the NATO countries and the countries of the Warsaw Pact has become a model of stability.

The United States has unequivocally aligned its national security interests with the interests of NATO, which in turn has been preoccupied primarily with the threat of an invasion from the East. In the words of Ronald Reagan, "Europe's shores are our shores. Europe's borders are our borders. We will stand with the Europeans in defense of our heritage of liberty and dignity."[2] However, going from such political pronouncements to agreement on deployment and control of military forces has been a giant and elusive step.

Europeans increasingly question the US commitment to stand and fight in Europe. Discussions of reductions of the American military presence in Europe raise their suspicions. Many other types of discussions between Soviet and American diplomats are also criticized as not giving adequate attention to the interests of NATO, and American political and military leaders are constantly flying to Brussels to explain that we have not sold out the alliance.

A personal experience illustrates the tensions within NATO that have persisted for many years as the United States works on several diplomatic fronts to promote its security interests. A number of years ago, I was asked by the Department of State to travel to Geneva and meet with the Soviet disarmament delegation considering a treaty to ban deployment of nuclear weapons on the seabed. We were to discuss some of the technical details of distinguishing between oceanographic instruments placed on the seabed for research purposes and nuclear weapons which might be clandestinely positioned on the seabed in violation of a treaty. The Department of Defense objected to such a discussion on behalf of the NATO Military Committee, which had not considered the issues, even though NATO had been informed that the discussions would be informal and would not address policy.

Thus, the Department of State changed my schedule, and I flew to Geneva via NATO Headquarters in Brussels where I first rehearsed my presentation before the Military Committee. After introductory remarks by the committee chairman exhorting the importance of intergovernmental consultation on all phases of the East-West confrontation, the members of the committee quickly lost interest in the slides that I had prepared to simplify my presentation, and many retreated to the coffee urn in the next room. I was simply going through the motions in Brussels to satisfy a political requirement. An attentive captain from the US Navy accompanied me to Brussels and Geneva since our navy was concerned that I, as a West Point graduate selected by the Department of State, might not adequately represent naval interests in the seabed. The navy then ensured the diplomatic propriety of all aspects of the trip by dutifully sending a report to the NATO Military Committee on the discussions in Geneva, emphasizing the "usefulness" of the prior consultations in Brussels.

Many Americans are skeptical of the commitment of the European countries to their own defense in view of the small share of their GNPs devoted to defense; Are the Europeans really paying their fair share of defending their own countries? these skeptics ask. Also, the US Government bridles at European trade policies which often appear to give the Soviets free access to high technology, while the United States protects the same technology in the interest of national security of the Western Alliance.

During my tour of duty with the US Army in Germany in the 1950s, I understood why, in the aftermath of World War II, the United States assumed the burden of defending that country while not allowing it to rearm and prepare to defend itself. Germany now makes major contributions to its own defense, although not in nuclear weaponry, yet the deployment of American conventional forces there has shrunk very little. The United States has always assumed that a large American military presence in Central Europe was politically wise; but changes in the East-West relationship and in West European politics require a reassessment by both the United States and the Europeans of the role and size of this American presence.[3]

The NATO approach to security has been summarized by the Department of State:

> NATO deters Soviet aggression through its strategy of flexible response and forward defense. Together they have proved remarkably successful; for the foreseeable future they continue to offer a degree of security which no alternative strategy could match. The essence of flexible response is to convince an aggressor that any attack on NATO—whatever its nature, place, or time—would expose him to incalculable and unacceptable risks. This requires NATO to demonstrate its political resolve to act jointly against all forms of aggression and to have the military capability to respond effectively at all levels of aggression. While this strategy does not require NATO to match the Warsaw Pact man for man or weapon for weapon, it does require the Alliance to maintain a mix of adequate and effective conventional and nuclear forces and to keep these up to date where necessary.[4]

This broad definition of the role of NATO provides maximum latitude in developing and deploying weapons systems. Many Soviet invasion schemes can be postulated. A popular perception is that NATO emphasizes preparations for responding to a Soviet armored strike across the heartland of Central Europe. The first line of defense would be the use of conventional forces. If they are overwhelmed, nuclear retaliation would be in order. While such a view greatly oversimplifies NATO doctrine, this theme remains at the core of NATO planning.

The Soviet view of NATO, at least the view which is articulated publicly, is quite different. The Soviets deny any aggressive intentions, and since 1987 the concept of "defensive defense" has become a cornerstone of the military doctrine which they state publicly. They say they are now trying to define conventional force requirements in Central Europe in terms of "reasonable sufficiency" to pursue this defensive strategy. In the absence of potential aggression from the East, they argue, NATO has itself become the aggressor; flexible response is simply a framework for preparing for unprovoked attacks against the Warsaw Pact. The Soviets point to the restructuring of their conventional forces in Europe, which is discussed below, as concrete evidence of their movement toward a purely defensive posture.

However, Soviet doctrine does not equate their defensive posture with a passive posture. They contend that they are drawing from the lessons of history in being committed to stopping an attack at or before the borders of their homeland. Once a battle begins, they intend to conduct the battle as fully as possible on the territory of the aggressor with the objective of terminating the aggressor's capability to continue to wage war. Such an approach does not accept the sanctity of national borders for protection of an attacker at the outset, during, or following armed hostilities. This philosophy signals a strategy which blurs defense and offense and which calls for a high degree of readiness at the very borders of the Warsaw Pact. Such a posture, together with the high state of readiness of NATO forces which also clouds the difference between offense and defense, is at the center of the problem of ensuring future stability of the frontier.

In most arms categories, the conventional forces available to the countries of the Warsaw Pact clearly exceed the conventional forces available to NATO. NATO's estimates project a ratio of tanks, other armored vehicles, antitank weapons, artillery, and air defense systems at about 3 to 1 and combat aircraft and helicopters at about 2 to 1 in favor of the Eastern bloc nations.[5] The Soviet estimates acknowledge Soviet undercounting of Warsaw Pact forces in the past, but still they present a different picture than NATO. Soviet estimates are summarized as follows:

. . . while the ground forces and air forces are roughly equal, the North Atlantic Treaty has a two-fold superiority over the Warsaw Pact in naval strength. The North Atlantic Alliance is superior to the Warsaw Pact in terms of the number of strike aircraft of front line (tactical) aviation and naval aviation, combat helicopters, and antitank missiles systems. The Warsaw Pact side has superiority in tanks, tactical missile launchers, air defense troop combat interceptor planes, infantry combat vehicles, armored personnel carriers, and artillery.[6]

Obviously the basis for counting differs. For example, the Soviets include the substantial French forces which have been withdrawn from NATO control, tanks and armored equipment assigned to reserve units, and American aircraft positioned on ships in the North Atlantic Ocean and Mediterranean Sea, whereas most of these forces are not included in the NATO estimates. Furthermore, reliance on "bean counting" without attention to the quality and capabilities of the equipment, which generally favor NATO, gives a misleading impression of the balance of forces.[7]

Despite the wide divergence in the estimates of forces, few military analysts believe that either side could overwhelm the other in a conventional war in Europe. Both East and West now seem convinced that the overall size of the conventional forces in Europe can be reduced. While the Soviets propose major reductions on both sides, NATO has thus far advocated only a 5 percent reduction in NATO forces if the Soviets reduce their forces drastically to reach a level of numerical parity.

Also, scholars from many countries are advocating reductions on an asymmetrical basis across different categories of weapons: this means that the Warsaw Pact would take deeper cuts in one or several categories such as tanks and armored personnel carriers, for example, in exchange for compensating cuts by the NATO countries in other categories such as aircraft. NATO has not accepted the concept, since according to its counting, the Warsaw Pact forces lead in every significant category; however, the approach will probably attract additional support in the years ahead. While the concept of trade-offs across weapons categories could allow each side to retain the mix of forces it considered most appropriate in military terms, translating the concept into arms control agreements will not be easy. The negotiations involve not just the superpowers but also the other members of NATO and the Warsaw Pact. Trading off the capabilities of different nations in one weapon category for reductions in the capabilities of other nations in another category in formal agreements raises very difficult problems for political leaders.

In a dramatic UN speech in late 1988, Gorbachev announced unilateral reductions of 500,000 Soviet troops stationed in Eastern Europe and the Far East and reductions of 10,000 tanks in Eastern Europe and the western military districts of the USSR. Of these totals, 5,000 tanks and 50,000 troops are to be removed from East Germany, Czechoslovakia, and Hungary. Also, artillery units in Eastern Europe are to be cut, and the capabilities of the USSR to deploy bridges which could support the influx of Soviet reinforcements across the rivers of Central Europe reduced.[8] His idea clearly is to demonstrate that the USSR does not plan to launch a blitzkrieg across Central Europe. Subsequently, several East European countries announced that they would also reduce the size of their conventional forces.

While the total size of the Soviet armed forces exceeds five million with over 50,000 tanks, the reductions scheduled during the next two years not only have great political appeal but are also militarily significant. In particular, these cuts may deny the Soviets the so-called "standing-start" option which has always bothered NATO. The Soviets will probably not be in a position to launch a surprise, conventional attack across Europe using only the troops and equipment deployed in Eastern Europe and sustain it until Soviet reinforcements arrive. Most importantly, the Soviets have indicated that they are prepared to reach agreements which substantially reduce conventional force levels in Europe while also redressing imbalances among categories of forces.

Will the United States and its NATO partners respond with their own unilateral reductions? The first Western reaction was cautious. "The Soviet cuts are a good start, but they should be deeper. They still retain an overwhelming conventional superiority. Will they allow international inspection of their reductions?" A few voices call for token NATO reductions as an encouragement for the Soviets to consider even deeper cuts. Formal arms control negotiations are now so complex and technology is advancing so rapidly that our diplomats spend most of their time negotiating problems of the past rather than of the future. Reciprocated unilateral reductions offer an alternative that deserves great attention.

Regardless of unilateral force reductions and the success of arms control agreements in reducing the military forces spread from the Urals to the Atlantic, the countries of NATO will continue to press hard to maintain technological superiority over the USSR on a broad front. NATO has always considered technology to be its trump card over the long haul in countering the preponderance of conventional forces of the Warsaw Pact. While Soviet weaponry in some armored and artillery categories may be more advanced due to heavier investments in specific technologies, NATO considers the Soviet weakness in the electronics field a serious flaw in Soviet efforts to stay abreast of Western advances. This technological edge for the West is bound to become even more important should the numbers of weapons decrease through arms control agreements. As would be expected, the Soviets in turn try to introduce constraints on further technological developments in their versions of arms control agreements.

Logically, cost savings should accompany reductions in conventional forces in Europe. This hope may be illusory, at least in the short run. Unless there are arms control limitations on upgrading existing weaponry, the demand for more sophisticated systems following an agreement limiting conventional forces will in all likelihood be so great that the actual expenditures on military forces in Europe will increase rather than decrease. Also, the costs of verification attendant to arms control agreements may be substantial.

Since the Soviets give extremely high priority to protecting their homeland, the possibility of nuclear attack is of special concern. Thus, the notion that NATO should have nuclear weapons in Europe is abhorrent to them, and the elimination of NATO's nuclear capability to reach the Warsaw Pact countries in general and the USSR in particular is a priority arms control objective. The Soviets will undoubtedly press hard to eliminate tactical nuclear weapons in Europe. This position already commands much sympathy within some of the countries of Western Europe. However, few Western political leaders will be prepared to abandon totally this counterweight, which they believe deters an attack by Soviet conventional forces in the absence of much larger reductions of these forces. In the longer term, as conventional forces are reduced substantially, elimination of tactical nuclear weapons seems like an achievable goal.

*   *   *

At the top of the list of Reagan's achievements during his eight-year presidency was the conclusion and ratification of the Intermediate Nuclear Force (INF) treaty. Few would have guessed that he would be the president to achieve the first agreement which actually reduces the number of nuclear weapons systems available to the two superpowers.[9]

Although the INF treaty eliminates all land-based missiles with ranges of 500 to 5500 kilometers that can carry nuclear warheads, it will be recorded in history as a modest military development. The reductions have limited military significance at this time; the large remaining arsenals of nuclear weapons include many strategic and tactical weapons that can cover the targets previously assigned to the INF systems. Similarly, violations of the agreement involving covert production of banned missiles, if there are any, will be inconsequential from the military perspective, at least in the absence of more far-reaching agreements to cover strategic and tactical nuclear weapons.

Politically, however, the INF treaty carries enormous significance. Most importantly, action has replaced rhetoric. The INF agreement has breathed new life into the disarmament process throughout the world. It has demonstrated that the two superpowers can be entrusted to reach agreements which do not undermine the interests of other countries. It has returned Soviet-American relations to a high level of rapprochement. The INF treaty shows that meaningful agreements to reduce weaponry, while complicated, are feasible. It has addressed the problems of verification directly and comprehensively and has established precedents for intrusive inspection considered impossible just a few years ago.

The agreement eliminates an entire class of missiles. Early in the negotiations each side was to retain a residual force of 100 warheads on missiles in this class, which would be deployed outside Europe. However, the marginal military value of such residual forces became apparent, and the impossibility of designing an inspection system for distinguishing between legitimate residual missiles and clandestinely produced missiles helped convince both sides to eliminate all the missiles. Also, the Asian countries, and particularly China, frowned on residual Soviet forces close to their borders, and both the United States and the USSR showed sensitivity to this concern.

The verification requirements for this agreement are quite complex and expensive. They loom important in setting precedents and providing confidence that both sides are living up to the agreement. Indeed, the arrangements are mind boggling. I was a participant in the early attempts of the US Government to develop on-site verification procedures for limitations on missile production in the 1960s. Then while serving as a diplomat in Moscow, I visited a few Soviet heavy industrial plants capable of supporting clandestine production. At that time, we would have been satisfied with one or two short-term inspections per year for ensuring compliance with agreements prohibiting or limiting missile production in such plants. Now, dozens of short-term inspections and continuing inspections by resident inspectors from the other country are in progress in the United States and the USSR.

Undoubtedly, technical violations will occur. Disagreements over the rights of inspectors will arise as they press the limits of their authority. Also, both sides will probably try to use inspection rights to gain intelligence information ancillary to the principal purposes of the verification provisions. Complaints will be raised at the political level providing opponents of arms control with grist for arguing against future agreements. Still, the precedents that have been established and the information on Soviet capabilities that will be gained will far outweigh such technical problems of little military consequence.

As might be expected, opinions are divided within the West on the next step after the INF treaty. The supporters of arms control and disarmament have clearly had their hands strengthened and want to build on the momentum from the INF success in reaching agreements on other reductions. They point to the current Strategic Arms Reduction Talks (START) and to the possibility of 50-percent cuts in the longer-range weapons systems including missiles and bombers. The more conservative forces want to let the dust settle, see how the Soviets perform under the INF treaty, and maybe in a few years take another modest step. Meanwhile, the Soviets are ready to consider significant cuts in many weapons systems, and they have their proposals for future reductions ready for our consideration at a moment's notice.

*   *   *

With regard to other types of limitations on armaments, restrictions on nuclear testing have long been at the center of the controversy in the United States over the desirability of far-reaching arms control measures. Arms control advocates contend that a ban on testing will place a technological cap on the nuclear arms race between the superpowers and will also encourage other countries, such as India and Pakistan, to refrain from the testing phase of nuclear weapons development. The weapons laboratories argue that testing is essential to develop warheads for new weapons systems which are being introduced and to test the continued reliability of proven warheads which are already deployed or in stockpiles. Environmental activists assert that nuclear testing even at great underground depths will inevitably release large amounts of radioactivity into the atmosphere. Others believe that the long-standing Soviet proposal for a ban on testing is only a ploy; once the United States has disbanded its effort, the Soviets will change their minds and again start testing, and the United States will be left behind. Who is right?[10]

Let me begin the debate from a personal perspective. As the senior official of the US Environmental Protection Agency (EPA) in southern Nevada from 1980 to 1985, I was responsible for protecting the population living in close proximity to the Nevada test site from exposure to radiation during underground tests conducted at the site. The population of primary concern was clustered in a few towns and in many small settlements, each with a handful of people, scattered along the test site periphery, which cuts through several hundred miles of barren desert. Before each test, which took place about every three weeks, either I or another EPA representative participated in the decision to detonate the nuclear device. We gave particular attention to the way it had been buried in the ground and to the predicted direction and speed of the wind which would spread radioactivity should an unanticipated leak from the test develop following detonation. During each test about 20 EPA specialists waited in pickup trucks just outside the boundaries of the test site to measure radiation levels and to assist with evacuation of the towns and settlements in the unlikely event of a radiation threat.

Only one incident disturbed my five-year watch. As I was delivering my daughter to her freshman dormitory in northern California, I heard on the car radio a news flash reporting a "major" leak at the test site. A quick telephone call to Las Vegas confirmed that a very small amount of radioactivity of little consequence had escaped. Following an underground detonation, the engineers drilled into an area hundreds of feet below the ground where the device had been buried. They were extracting "hot samples" for analysis; unfortunately, a small seep developed as they punctured successive geological layers.

The interest of the press and television grew intense. The groups opposed to nuclear testing, which frequently manned the picket lines at the gate of the test site, were spreading exaggerated stories of the environmental dangers. They clearly wanted to use this event to discredit the testing program.

Upon my immediate return to Las Vegas, our EPA monitoring specialists determined that up to 100 residents of a small town on the edge of the test site could have been exposed to radiation from the seep. Had they remained outside their homes for several hours as the radiation plume passed through the town, their maximum exposure to radiation would have been equivalent to the radiation exposure they would receive while standing for three minutes in the southern Nevada sunshine. And this was the most "serious" radiation release in a decade. In short, objections to underground nuclear testing should not be based on environmental arguments since we know how to conduct tests without threatening the residents living nearby. We also know how to protect the workers closer in at the site of the tests. My American colleagues who have been to the main Soviet test site at Semipalatinsk tell me that while the Soviets are less concerned with environmental research, they know how to bury devices at great depths to contain harmful radiation.

However, as I waited quietly in the command center of the test site at five o'clock in the morning in anticipation of each test, I often had serious reservations about the necessity for the tests—about their contribution to our national security, about their high costs, and about their diversion of highly talented specialists from other pursuits. Feeling the ground shake at detonation time and watching the TV screens record the sinking of the ground surface above the test hole a few minutes later certainly impressed upon me the power of technology. At one time there were high hopes that nuclear explosions could be used for peaceful purposes—for large excavation projects, for tapping into buried natural resources, and for reshaping the landscape. These hopes have largely faded, and this explosive power now has no use other than destruction—destruction of people, of cities, and of civilizations.

Many arguments can be made in support of a continuation of underground nuclear testing. We need to confirm that our weapons which have been in stockpiles for many years continue to work properly. We need to test new warhead configurations which are lighter, more reliable, or more efficient. We need to test designs for new weapons systems. We need to understand the effects of nuclear detonations on communications systems and on materials used in military systems. We need to permit the scientists from our weapons laboratories to test their concepts, or they will grow stale or seek other employment. We need to test so we can better understand the significance of Soviet tests. We need to test to show the Soviets that we mean business.

The arguments against nuclear testing are equally persuasive. The United States has conducted more than 700 nuclear tests, some involving several hundred experiments; and information from additional testing is of decreasing importance. Data from earlier tests have not been fully utilized, since fresh data are more interesting than old data. Weapons could be designed with greater attention to long-term reliability so that proof testing is far less important than in the past. Computer simulation of testing has advanced to the stage where we can predict performance of weapons with increasing certainty. Testing of many components can be carried out without detonating the nuclear device. Periodic testing to check the reliability of weapons already in our inventory has little technical merit: the number of weapons of any single type that can be tested is so small that conclusions that are statistically valid about the continued reliability of the other weapons of the same type are not possible. Finally, tests provoke militaristic responses by the Soviets.

American military strategy rests on deterrence. The United States does not plan to use nuclear weapons. If the Soviets strike first, the United States is in a position to respond with more than enough firepower to inflict massive destruction. If both sides reduce their nuclear stockpiles, both sides will be more reluctant to launch a first strike, particularly if there are uncertainties about the performance of the nuclear weapons in destroying the other side's retaliatory capability.

Thus, it doesn't really matter whether we can predict with precision the explosive power of nuclear weapons. It doesn't matter whether they are the most advanced configurations. There is no doubt in anyone's mind that they will work, and they will wreak enormous devastation. With a heavily armed world, they will serve their purpose of preventing war. As the arms levels decline, the remaining nuclear weapons will continue to raise serious doubts about the wisdom of a nuclear first strike or a conventional attack.

In 1963, the United States and the USSR agreed to end testing in the atmosphere, and they also assumed a treaty obligation to work toward a complete ban on underground nuclear tests. During the dozen years that followed, the commitment to seek a ban on underground testing received low priority in both countries, and underground testing continued at a brisk pace. Finally in 1974, the representatives of the two countries agreed to prohibit testing of nuclear weapons with an explosive power of more than 150 kilotons—a threshold for weapons seven times more powerful than those used in Japan during World War II. While this limitation was included in a signed agreement, the US Senate has still not ratified the document. The two countries have generally refrained from testing larger weapons for the last 15 years. However, the United States has charged that the Soviets have conducted some tests with yields exceeding 150 kilotons. But some scientists in the United States as well as in the USSR question this assertion. They doubt the reliability of the techniques used by our government in relating the size of the Soviet underground explosions in unfamiliar geological formations to the seismic signals that are generated and then analyzed as the basis for the estimates.

Despite acceptance of the 150 kiloton cap, the US record in living up to the commitment made 25 years ago to work for a total ban on all underground testing has not been very good.

The United States has not matched the Soviet record, which most recently included a unilateral moratorium on testing for more than 18 months. During the past few years the United States has repeatedly cited the lack of technical verification capabilities as a principal reason for holding back on further limitations on nuclear testing.

As previously noted, the principal reasons put forth by the United States for continuation of an active test program have shifted to military requirements to "modernize" nuclear weapons and to test the weapons in our inventory to make sure they work as they become older. The nuclear weapons community in the United States speaks with a powerful, unified voice which resists limitations on testing. It is a closely knit group of highly talented and dedicated specialists with a proud record of major technological achievements and a tradition of generous budgetary resources. Indeed, this community controls 60 percent of the budget of the Department of Energy as well as enormous resources within the Department of Defense.

A total ban on underground testing does not seem likely in the near future. The argument that a policy of nuclear deterrence requires some level of testing to ensure that the deterrent works is politically powerful. Also, the SDI program includes a nuclear testing component, and this program will undoubtedly continue albeit in a form other than originally envisaged.

Assuming that the United States and USSR continue their active search for areas of agreement in the field of arms control, new limitations on the yield of underground tests seem to be a reasonable near-term expectation. Reducing the yield limitation on permissible tests from 150 to 10 kilotons, which would allow testing of some key weapon components, probably will eventually be accepted by the advocates of continued testing, but they will strongly resist reducing this limitation further.

As to the number of tests, there is currently no limitation. The number of announced tests by the United States is between 15 and 20 each year. Occasionally, unannounced tests are carried out. A limit of 6 to 8 tests per year should be a feasible near-term target. Such a level of testing will allow each country to continue its most important programs and to be in a position to resume full-scale testing if the international situation deteriorates to the point of withdrawing from treaty obligations. Unfortunately, such a reduction will not cut costs significantly since the logistics associated with preparing and conducting tests will not decline in a major way. Indeed, the costs of testing may rise as the number of experiments attached to a single test will most likely increase dramatically to compensate for fewer tests. If the budget for testing is preserved, the financial beneficiaries of the testing program will probably not object so strenuously as to block modest limitations on the number of tests.

Developing a technical rationale for a quota on the number of tests will be difficult. A popular suggestion is to allow enough tests to check on the reliability of weapons in the stockpile but not enough to permit development of new weapons. Remembering that hundreds of experiments can be carried out during a single test and recognizing the difficulty of verifying whether an adversary is testing an old or new nuclear device, such a technical argument cannot be sustained. Nevertheless, limitations on the number of tests can be politically important as a component of an overall arms control approach and as a step toward an eventual comprehensive test ban.

*   *   *

Turning directly to SDI, shortly after Reagan's announcement in 1983 of the SDI initiative, I was waiting for the next underground test at the Nevada test site together with scientists from Los Alamos National Laboratory, one of the American laboratories which designs nuclear weapons. At that time, the most illustrious alumnus of this weapons laboratory was Jay Keyworth, the former presidential science advisor and a laser specialist who played an important role in developing the SDI concept for the president. Naturally the conversation turned to this new presidential initiative, which meant a great budget increase for the Los Alamos laboratory. All of us knew from firsthand experience the destructive power of nuclear weapons, and we were eager to see the development of new systems which would prevent nuclear weapons from ever impacting on US territory. But could a defensive system be designed that would not be overwhelmed by offensive systems?

Keyworth's colleagues from his old laboratory were suddenly confronted by a real dilemma. They couldn't be happier with the news that their laser research program and many related activities would be generously funded for the indefinite future. They were deeply committed to exploring the potential of laser systems for improving strategic weaponry. At the same time, they simply could not accept the technical arguments being advanced in support of the overall SDI concept. They shared the views of some of the harshest SDI critics over the impossibility of ever deploying an effective system at anything less than astronomical costs and, even then, having only a system which could be overwhelmed by new offensive systems. The best solution to their immediate dilemma was a retreat to the bar and a large bottle of Green Hungarian wine, a popular off-duty beverage at the test site. This wine quickly helped all of us forget about the world of technical reality while my friends from Los Alamos concentrated on ways to spend their new research budgets.

Meanwhile, in Moscow the initial wave of Soviet criticism of this "provocative" act on the part of the United States was unleashed, and Soviet diplomats suspended bilateral talks on cooperation in peaceful space exploration. The immediate Soviet reaction was to challenge the US contention that SDI was strictly a defensive system, arguing that the real purpose of the program was to enhance US first-strike capabilities. Soviet scientists were soon hard at work trying to sort out the technical feasibility of SDI. They took the position, at least publicly, that the technical problems associated with the construction of an electronic astrodome the size of the United States were enormous and probably could never be solved. Their confidence in their assessments was boosted enormously by the many technical arguments presented by American scientists who spoke out against the SDI program.[11]

According to stories circulating in Moscow, a few young Soviet scientists were stimulated by this new concept of an impenetrable shield and were eager to see if they could contribute to promoting the concept in the USSR. They apparently were soon dissuaded by senior Soviet scientists who convinced them that, even though Soviet military scientists were exploring this technology, the concept was not particularly promising and they should channel their energies elsewhere.

The SDI program greatly bothers the USSR. First, the Soviet leadership believes that even if the motivation for SDI is purely defensive, many subsystems developed within the SDI program will have offensive applications—particularly systems related to communications, to keeping abreast of activities on the battlefield, and to targeting warheads. Also, the Soviets are not interested in transferring the potential battlefield to space, where they will have a serious technological disadvantage due to their general backwardness in the electronics field. Finally, one result of the SDI program will probably be a widening of the already large gap between the two countries in electronics.

The Soviet response to SDI depends on technical and political judgments. Initially, they seemed to overreact to the likelihood that SDI might soon be on line. Later they became skeptical that the United States would sustain the effort in the face of domestic criticism over technical feasibility. By now, though, they have seen the program develop through five years of the Reagan administration, and they are undoubtedly convinced that the program will continue in some form.

The Antiballistic Missile (ABM) treaty signed by both countries during the 1970s limits testing of antimissile systems and gives the USSR a powerful argument against testing of SDI by the United States. However, the US Government contends that some testing of SDI is permitted under the treaty. The policy debates become further complicated in view of the Soviet testing of their older ground-based antimissile system which is permitted under the treaty.

The Soviets are aware of the progress and difficulties in their own air defense systems and in development of new antimissile systems, including components related to some elements of the SDI concept. The high costs of their efforts to date will only increase as they move from research to development and perhaps eventually to testing and deployment. Thus, the depressed state of the Soviet economy and the slow pace of Soviet technological development clearly add to their concerns.

For the time being it appears that the Soviets will continue with development and testing of their own antimissile systems but will aggressively pursue strengthening of the existing treaty and clarifying the grey areas of disagreement to limit testing or deployment in space of any type of offensive or defensive weapons systems. If the arms control route proves fruitless in slowing down the development of the SDI system, Gorbachev may have to respond to internal pressures from the Soviet military establishment in a more politically dramatic manner to retain the Soviet image as a major power in the face of the SDI program.

This response would probably emphasize new offensive systems which could confuse detection capabilities and find holes in the SDI shield.

A particularly undesirable scenario could follow failure on the arms control front. The Soviets might respond to SDI with a new generation of offensive strategic weapons. Meanwhile, lack of strong public support in the United States for an uncertain defensive system, coupled with technical difficulties, could result in abandonment of SDI. The United States would then be faced with a new Soviet offensive threat. A likely US response to domestic pressures to match the Soviets step by step would be development and deployment of new offensive systems of our own, drawing on previously developed SDI technology to the extent possible. This would be a sad concluding chapter to this phase of military history.

*   *   *

For many years the Western powers have assumed their technological edge over the Soviets in military weaponry to be their strongest asset for ensuring our security. The United States in particular has successfully developed and deployed many critical strategic systems well in advance of the USSR. The early capability of the United States to build quiet submarines, to use solid rocket fuels, to place multiple warheads on single missiles, and to deploy cruise missiles, for example, has provided important temporary advantages over the Soviets. Now our government looks to SDI and the stealth bomber as additional technological advantages which it must exploit to guarantee national security.

In each case, however, the technical edge has been temporary, as the Soviets have continued their relentless buildup of military systems. Generally, they have relied on the brute force of large numbers of less advanced systems to compensate for technological shortfalls until they are able to emulate the American systems. But each innovation on the part of the United States has added new impetus to the arms race, and eventually the advantage of any single innovation has disappeared as the Soviets have mastered the technology or have developed counter measures. The United States in turn has searched for additional technological innovations that would raise the arms competition to a still higher level.[12]

The vicious cycle of American innovation, Soviet response, another American innovation to thwart the response, another response, and so forth must be broken. It is both costly and dangerous. A 10-year technological lead has lost much of its significance, since the technology of the 1970s can still deliver a potent punch. Numerical advantages have also lost much of their meaning in view of the large numbers of weapons available to both sides. Mutual restraint in military spending together with formal arms control agreements and reciprocated unilateral arms reductions are the essential ingredients in putting aside the notion that there can be permanently meaningful technological or numerical superiority when it comes to nuclear weapons.

*   *   *

In summary, Soviet and American strategic assessments increasingly recognize that military stability in the superpower relationship can be maintained with much lower levels of nuclear and conventional forces on each side. Recent history has vividly shown that neither side has been able to gain a decisive edge through expansion of its military forces. Now the challenge is for both sides to take actions that will indeed reduce the military confrontation to a much lower level of military forces. Eventual reduction of the nuclear force levels of both countries by 75 percent and substantial cuts of conventional force levels on both sides should not frighten even the conservative elements of the two societies.[13]

Placing a qualitative limit on the arms race will be more difficult as each side seeks the most advanced systems, lest the other gain a technological advantage which will be decisive. Limitations on testing of new systems, together with limitations on military budgets, are one approach to slowing the technological race that needs much greater attention.

Techno-diplomacy is the art of the possible. We should work toward realistic near-term objectives while retaining the goal of more substantial reductions of force levels in the longer term. The two countries should promptly agree to 50-percent reductions of strategic nuclear weapons systems, more stringent limitations on underground nuclear testing, and the first steps in mutual reductions of conventional forces in Europe. These steps will not alter the current balance of power. They can significantly change the nature of the East-West confrontation in a direction which gives a greater priority to peace. While many technical details need to be worked out, agreement in each of these areas seems clearly possible with only a modest amount of political will.

Also, the two countries should rely more and more on unilateral actions to reduce their armaments while awaiting the completion of complicated formal agreements to codify such actions. As the serious dialogues on military matters between the two countries continue to increase, the specific opportunities for such actions which are in the interests of both countries should emerge.[14]

CHAPTER 3

Soviet Science and Technology Fall Behind

*Soviet research workers and engineers resemble
soldiers attempting to fight a modern war with
crossbows.*
Soviet scientist Roald Sagdeyev

"Plagued by low productivity, lax standards, and bureaucratic stultification, Soviet science needs urgent attention and reform." These words of Gorbachev's adviser Academician Roald Sagdeyev were published in the summer of 1988 in the American journal *Issues in Science and Technology*. Earlier that year Sagdeyev had leveled similar criticism at the Soviet scientific enterprise in the Moscow daily *Izvestiya*. Sagdeyev reflects the accumulated frustrations of many Soviet scientists in their efforts to unleash the potential of the Soviet science and technology community.[1]

Sagdeyev, who is a leading space scientist, continues, "For too long, Soviet science has hidden its inadequacies behind official panegyrics to its success. In academic and political forums alike, exaggerated claims have been made for the achievements of Soviet science. Science has its own criteria for success, however, and Soviet achievements have not measured up to them."

Sagdeyev urges a major revamping of the Soviet approach to science, including breaking up the massive national research institutes he describes as "bureaucratic dinosaurs" into smaller research groups, declassifying much of the research now considered to have military importance, and relaxing limitations on international scientific cooperation.

"Soviet science has suffered deep, and still bleeding, wounds from ill-conceived government policies," he states. He is particularly disturbed over restrictions on the mobility of Soviet scientists, both geographically and professionally, due to the "rigidity and compartmentalization of the research bureaucracy" and an acute shortage of housing in Moscow, where much of Soviet science is concentrated. He also bemoans the use of antiquated equipment, especially the primitive computers, in Soviet laboratories.

Sagdeyev is not alone in his criticism of Soviet science. From sharp rebukes by Gorbachev at the top to the complaints of science students at the universities, Soviet technological achievements and lack of achievements in comparison with accomplishments abroad are under constant attack. The Soviet media are having a field day in publishing exposés of mismanagement of scientific institutions—institutions previously shielded from public scrutiny and criticism. Economists are particularly eager to jump into the fray and to try to demonstrate that research must be driven by economic principles concerning inputs and outputs.

Who is at fault? No one or everyone? Industrialists blame the poor quality of their products and their inefficient production techniques on the lack of support from the scientific and engineering communities. Soviet scientists blame each other, blame the government bureaucrats who allocate resources, but most of all blame the centralized command and control system which manages all facets of economic and scientific endeavor.

All the while, Soviet spacecraft circle the earth and explore the planets, the Soviet military boasts many types of modern weaponry, and the Moscow subway continues to amaze foreign visitors. However, the Soviet public is tired of hearing about these achievements. The public simply cannot understand how a nation which has over one-half of the world's engineers and one-quarter of the world's scientists cannot solve its basic problems of food, housing, and health care. The Soviet scientific community is equally unimpressed by these accomplishments of the past. It will never accept the Lenin Prizes for scientific achievements as an adequate substitute for Nobel Prizes, which have eluded Soviet scientists for many years.

*   *   *

The roots of Soviet science go back more than 260 years, but they developed slowly. In 1724, Peter the Great established the Academy of Sciences in St. Petersburg (Leningrad). One purpose of this new institution was to attract foreign scientists to the USSR. In the generations that followed, this window to the West widened. Many prominent European scientists discovered Russia, often out of curiosity, and a number of Russian scientists gradually earned worldwide recognition. In particular, Pavlov's experiments on responses to physiological stimuli and Mendeleyev's ordering of the properties of chemical elements stirred great interest in the West. By the beginning of the twentieth century, the Academy in St. Petersburg had established itself as a well-regarded institution destined to play a significant role in the future evolution of the Russian and then the Soviet societies.

The importance of technology, or lack of technology, gained recognition during the battles on the Russian front in

World War I. The new Soviet leadership promptly took steps to enlist the support of the scientific community. Following the war, the system of central direction of science and technology seemed appropriate to the needs of the new Soviet state. Indeed, within several decades the Soviet Union changed from a nation with a very limited scientific effort to a nation with one of the world's largest scientific establishments.

In the 1920s and 1930s the USSR did not hesitate to adopt Western engineering approaches and invited hundreds of American and other Western engineers to assist in developing the country. The Ford Motor Company, for example, provided essential technical expertise for the construction of the Volga automobile plant in the city of Gorkiy on the Volga River, and several of the large Soviet hydroelectric plants trace their origins to technologies and engineers from the United States.

Meanwhile, within a few years after the 1917 revolution and during the period of recovery from World War I, Lenin accepted the recommendations of his advisers to diverge temporarily from the road to socialism in deference to an urgent need to stimulate economic growth. He, like Gorbachev 65 years later, decided to allow a private sector to develop on a limited scale: private entrepreneurs could have small factories which employed up to 20 employees. In view of the critical food shortages in the early 1920s, the peasants could keep and sell some of their products on the free market after paying a tax to the state. At the same time, Lenin tried to require the large state enterprises to account for their costs and to operate in the black.

During this brief period in the 1920s when Lenin's New Economic Policy was in place, the Soviet economy gradually revived, and industry reached prewar production levels. The private sector clearly made a contribution to this revival. Retail trade was dominated by private entrepreneurs. A middle class of wealthy businessmen, who were called Nepmen, and wealthy peasants, the kulaks, emerged. This brief interlude ended abruptly with the death of Lenin in 1924 and the ascendancy of Stalin. A rigid central planning system was instituted, terminating the flirtation with private enterprise. Nostalgia of the past has recently come to life in the Soviet ballet *Golden Age* which brings to the stage of the Bolshoi Theater the café life of the wealthy class during the 1920s.[2]

Returning to science and technology, we note that the Soviet approach has always been greatly influenced by the size, climate, resource distribution, and demographic character of the country. As would be expected, Soviet science and technology are also very much a product of the political and economic systems of the Soviet state. However, the wide diversity in the geographic and demographic factors of this sprawling country have seriously complicated attempts to simplify approaches through central planning and control. Over the years many research institutions in outlying areas of the country operated as islands unto themselves, with communications between Moscow and distant program managers and scientific colleagues limited and, at times, nonexistent. Senior scientists in some instances were simply given blank checks to manage their institutes as they saw fit for many years. Even in the case of the country's leading mathematics institute in Moscow, the director remained in place for 50 years until his recent retirement.

World War II provided a great incentive for the Soviet scientific community to pull together. Considerable energy was devoted to the development of nuclear weaponry; fission and then fusion were mastered shortly after the war. This successful effort encouraged large increases in allocations of resources to science and enhanced the standing of Soviet science at home and abroad. During the 1950s and 1960s several Soviet scientists were awarded Nobel Prizes. Relatively narrow Soviet accomplishments in military technology, Soviet pioneering achievements in space exploration, and these early Nobel Prizes contributed to an exaggerated view of Soviet science in the West.

Such dramatic Soviet scientific achievements partially overshadowed the research philosophies propounded by a scientist who became the leader of the Soviet agricultural research effort, Trofim Lysenko. He managed to convince the Soviet political leadership that the realities of genetic inheritance were grossly distorted at home and abroad. Consistent with the Marxian writings on environmental determinism, the genetic characteristics of plants and animals could be modified by environmental factors, he argued. In the late 1940s, world-renowned Soviet geneticists were displaced, and laboratory research and field-testing programs were redirected to prove misguided theories. Soviet agricultural research entered a period of decline from which it has yet to recover fully.

When Khrushchev appeared on the scene following Stalin's death in 1953, he instituted a variety of management changes to galvanize Soviet industry and agriculture in pursuit of his long-term goal of overtaking the United States. He introduced many organizational changes, including some decentralization of management of industrial activities. However, most administrators and planners remained firmly in place, and the changes were not dramatic. Agricultural and industrial productivity still lagged far below Western averages.

One of Khrushchev's most notable achievements with regard to science was to encourage the establishment of a scientific base near the Siberian city of Novosibirsk as part of his effort to exploit the potential of the vast lands to the east of the Ural Mountains. He supported new scientific research institutes there and throughout Siberia to coordinate activities from the Urals to the Pacific. The young scientists who moved to Siberia to start their own laboratories were very proud to show me and other early visitors to the region 25 years ago the world's most powerful research cannon for beaming jets of water at metallic targets to test the properties of the metals, the most advanced accelerator which used the principle of colliding beams of nuclear particles, and the world's leading center for biological research on sables.

Despite all of this, economic development of Siberia moved forward only slowly. Although western Siberia currently produces almost two-thirds of Soviet oil and gas and the industrial cities sprinkled across central Siberia continue to grow in importance, Siberia remains a vast undeveloped area with transportation and communication networks still concentrated along a few important arteries.

The isolation of Siberia hit home in 1965 when my wife and I first visited Yakutsk, a city in north central Siberia accessible only by a plane flight of several hours from Irkutsk and, in the summer, also by a boat trip over several days on the Lena River. But even in this remote setting science was alive. I visited the log cabins clustered together in a medical school where open-heart surgery was performed. I watched the launching of meteorological balloons and then followed the tracking from the dome of an old church which had been converted to a research laboratory. And like every scientific visitor to this city, I wandered through the underground ice caverns to observe research on construction techniques in permafrost regions. Other more recent visitors to Yakutsk report that research continues in these and other directions.

During the Brezhnev era, which lasted from the mid-1960s until his death in 1982, funding for scientific research continued to grow. The research effort put new emphasis on the economic payoff from research. Introducing the results of research into practice became a theme which began to dominate all discussions of research priorities and research funding. Whenever a new technical problem arose, a new research laboratory was organized to solve the problem and then to turn the solution over to a ministry or industrial enterprise for implementation. However, researchers found few customers for their ideas or their new designs since the ministers and the enterprise managers were fully occupied with meeting production quotas. They were not motivated to disrupt work to try out unproven concepts.

The British Scientific Counselor in Moscow recently described the Brezhnev era of science and scientists as follows: ". . . a great ocean liner drifting rudderless on the open sea with a third of the complement beavering away, happy in the knowledge that they were busy, a third keeping their heads down and enjoying some peace and quiet, and a third controlling what the other two thirds were doing."[3]

*   *   *

Thus, a stage bedecked with talented but frequently uninspired scientists and engineers was set for the arrival of Gorbachev. In the eyes of Soviet scientists, here at last was a man of vision, daring, and determination. They could relate to him and he to them. But in 1986 his initial confidence in Soviet science and technology was jolted by the tragedy at the nuclear power plant in Chernobyl. Nuclear power was to be a keystone in the development of the USSR; it would overcome the dilemma posed by the remote location in Siberia of oil, gas, and coal reserves and the ever-increasing demand for energy thousands of miles away in the European part of the USSR. The accident vividly brought to light the weaknesses in Soviet engineering practices and in the training and management of personnel responsible for modern technological systems.

Within several hours of the accident, Academician Valeriy Legasov, a highly respected nuclear scientist, was selected by the Soviet authorities to play a leading role in the cleanup and investigation. Also, he represented the USSR in international discussions of the events that transpired. Two years after the accident in May 1988, at the age of 52, he committed suicide, perhaps in a state of despondency over this tragedy. Just prior to his death he prepared notes on the accident, the events that led up to the accident, and the response to the accident for publication in *Pravda*. The following excerpt from these notes is a telling commentary on the Soviet approach to an important area of science.

> After being in Chernobyl, I drew the unequivocal conclusion that the Chernobyl accident was the apotheosis, the summit, of all the incorrect aspects of the running of the economy which had existed in our country for many decades. There are not abstract, but specific, culprits for what happened at Chernobyl, of course. We now know that the reactor protection control system was defective, and proposals had been made to eliminate this defect. Not wishing to become involved in additional work, the designer was in no hurry to change the protection control system. What happened at the Chernobyl power station had been going on for a number of years. Experiments were carried out in accordance with a program which had been drawn up in an extremely negligent and sloppy manner. There were no rehearsals of possible scenarios before the experiments were conducted. . . . The disregard for the viewpoints of the designer and scientific leader was total, and the correct fulfillment of all the technological procedures was a struggle. No attention was paid to the state of the instruments or the state of equipment before it was time for planned preventive maintenance. One station director actually said, "What are we worried about? A nuclear reactor is only a samovar; it's much simpler than a conventional station. We have experienced personnel, and nothing will ever happen."[4]

Gorbachev's perspective on Soviet science received a second jolt from the earthquake in Armenia in December 1988. Poorly constructed buildings collapsed, greatly magnifying the death toll, which eventually exceeded 25,000 people. Even after the devastating earthquakes in the Soviet Central Asian cities of Ashkhabad in 1948 and Tashkent in 1966, Soviet construction authorities had ignored acceptable building practices; and serious inspections and rejections of newly constructed buildings remained virtually unknown in the USSR.

Construction officials throughout the country are under great pressure to erect buildings fast—to house a burgeoning population. In Armenia they apparently determined that they didn't have time or adequate materials to prepare for the most serious earthquake ever to erupt in a populated part of the region. In 1972 the height limitation on buildings was raised from five to nine stories, despite the known seismicity of the region, in order to accommodate more people quickly; many of these buildings were subsequently destroyed. Prefabricated concrete segments of buildings were not tied together properly; they collapsed like decks of cards. Also, the quality of concrete was poor; in some cases structures pulverized during the shock.[5]

Twenty-five years ago I was shown detailed Soviet maps of the earthquake-prone regions of the country. The cities were carefully zoned to account for the small variations in the likely seismic effects on structures due to the geology of each region. Building authorities boasted that future construction would respond to these careful scientific investigations. If only they had followed through on this boasting.

*   *   *

The annual Soviet investment in research and development (R & D), including both military and civilian programs, is on the order of 70 percent of the US investment, even though the Soviet GNP is one-half of the US GNP. A reliable breakdown of Soviet expenditures between military and civilian R & D is not available, but the Soviet Union clearly devotes a much higher percentage of its R & D investments to military systems than does the United States. The Soviet investments in military R & D probably approach US investments. Regardless of uncertainties in budget allocations, the Soviet R & D effort is very large. This high level of expenditure is particularly striking since in many fields the Soviet Union lags behind Western countries with far more modest research expenditures.[6]

Soviet research is dispersed throughout several thousand scientific institutes and laboratories. The most important activities are concentrated in Moscow, Leningrad, Kiev, and Novosibirsk and in several satellite science cities near Moscow. In addition, most of the capital cities of the outlying republics have sections dedicated to scientific institutes and experimental facilities. The staffs of the institutions in these cities are very gradually moving into housing in nearby residential areas as new construction is completed. The clustering of scientists in science cities and science districts of large cities reduces commuting problems and permits employers to provide cultural and social services as well as professional opportunities. However, clustering also contributes to the isolation of science from practical reality, isolation that has been the target of criticism not only in the USSR but in other countries as well.

Many field-testing programs are concentrated near these scientific centers. They also take place in other cities where heavy industrial activity is located, particularly in central Russia, the Ukraine, the Baltic republics, and western Siberia. Military testing of weapons systems and military detection devices is also conducted in more remote areas of Siberia, Central Asia, and the far north.

The Academy of Sciences of the USSR is the most important scientific institution in the country. Since moving its headquarters to Moscow in 1933, the Academy has played an increasingly important role both in directing the national effort in fundamental research and in providing advice to the government through its individual members and through a network of standing and ad hoc committees. Among its 900 members are many of the most influential scientists of the country. Many members hold key positions in the government, and others manage large scientific and engineering complexes. A few hold senior posts in the Communist party, and they influence the general policies guiding Soviet science. Over 150,000 scientists and support personnel work in the 250 institutes of the Academy. Offering relatively good working conditions and considerable prestige, the Academy attracts many of the best young scientists of the country. The Academy also exerts considerable influence over the programs of another 300 institutes which belong to the Academies of Sciences of the fourteen outlying republics of the USSR. These institutes are generally not as strong as the central institutes, although a few, such as a welding research institute in Kiev, have achieved international stature.

The Academy receives only a modest portion of the nation's resources devoted to R & D, less than 10 percent. Its share includes about two-thirds of the resources devoted to fundamental research removed from applications, which is far less costly than development. The bulk of the nation's R & D resources are earmarked for technology and go to the many other institutions conducting applied R & D in direct support of the industrial ministries. A very limited amount of research support is provided by the government to the universities and other institutions of higher education. Research resources are also devoted to the agricultural and health sectors, largely through the separate Academies of Agricultural and Medical Sciences.

Soviet research seems to cover almost every aspect of science. The principal exceptions are those research areas which depend on very high speed computers or on sophisticated experimental facilities that are not available to researchers in the USSR. There is hardly a topic of scientific interest which is not being studied by a Soviet scientist somewhere in the system. Frequently when a new product appears in the West, Soviet researchers immediately initiate a project to develop a Soviet analog, even though they may already be three to five years behind when they begin their project.

Heavy concentrations of Soviet scientists focus on specific research areas which have been determined to be of high priority by the government—genetic engineering, polymer chemistry, nuclear physics, to name a few. Indeed, no other country can match the Soviet capability for mobilizing scientific manpower to address selected topics. Most research programs have been determined by top-down planning; only now are individual Soviet researchers being encouraged to come up with their own ideas as to research areas of greatest promise.

As to the quality of Soviet scientific research, a good descriptor is "uneven," with fewer bright spots than dark spots. The best Soviet theorists equal their Western colleagues in some fields of science. They have made pioneering contributions in mathematics, physics, chemistry, and the earth sciences, for example. More specifically with regard to physics, their current work in laser physics, dense plasmas, and fusion rank at the top of the world level. I recently heard the president of the Academy of Sciences of the USSR state that his experts estimate that Soviet achievements in basic science are at a world level in about 50 percent of all scientific fields; however, such estimates seem high, based on reports of many Western scientists who have visited research facilities throughout the USSR.

While many Soviet experimentalists are extremely talented, they are greatly hampered by the lack of equipment and supplies, as previously mentioned, and they lag behind the West in most experimental areas. Exceptions are noted, primarily in areas which are not dependent on sophisticated electronic techniques. Moving closer to applications, Soviet efforts fall far behind. In Moscow I learned that British experts have a very low opinion of Soviet applied research: they classify only 10 percent of the applied research activities supporting the 24 principal branches of Soviet industry as world-level technology.

Soviet facilities which house the civilian research effort are generally poor. Most of the buildings are too small, frequently in a state of disrepair. Laboratory space is at a premium. Heat, power, gas, and water can be unreliable. There are exceptions: for example, the Institute of Bio-organic Chemistry in Moscow is an outstanding modern facility as the result of heavy invest-ments, including $75,000,000 in Western currency for the pur-chase of foreign equipment. I, like other visitors to the Institute, was particularly intrigued by the layout of the buildings, which when viewed from above simulate the structure of a compli-cated molecule. As another example, the Space Research In-stitute in Moscow is sufficiently well equipped to be a hub for planetary exploration. Modern nuclear research facilities can also be found in several cities.

Despite the harsh criticisms currently being directed at So-viet research by both Soviet and foreign scientists, the severe lack of hard currency for research, and other constraints sur-rounding Soviet scientists, the Soviets have made and continue to make many important contributions to international science. Even in those fields where an overall assessment will clearly show a lag of many years behind Western colleagues, Soviet scientists are laboring in narrow subfields at the frontiers of science. These activities reflect much of the 50 percent of Soviet scientific excellence referred to by the Academy president.

Soviet science has occasionally profited from its depriva-tions. As the Western world rapidly changes its approaches to research in response to new technological developments, Soviet institutions often stick to the older, unsolved problems. For ex-ample, the Soviet effort in studying agricultural viruses, which is a labor-intensive research activity, has achieved excellent results. Also, for many years the Soviet research effort has been forced to rely on long-term approaches using proven experimental technologies. The most dramatic example is the consistent use of standardized space rocketry and instrumentation. These standard approaches have high reliability, and scientists can usually predict with assurance the types of results that are likely to ensue from experiments. The Soviets have also developed a high level of craftsmanship in fashioning experimental equipment using only those materials available in the USSR— for example, electronic measuring devices, chemistry glassware, and metal-testing equipment. On occasion Soviet scientists have successfully improvised to stay abreast in fields which are driven by computer technology in the West, developing alternative approaches and programs that rely more on "brainpower" than computer power.

In addition to shortages of space, facilities, and equipment, isolation fostered by the closed nature of Soviet society has had a stifling effect on more rapid advancement of Soviet science. Soviet scientists are isolated from colleagues abroad working on similar problems. Very few Soviet scientists travel, only a limited number of laboratories host visitors from abroad, and foreign scientific journals are a scarce commodity everywhere. As already noted, the shortage of copy machines and the high level of control over their use further limit attempts to exchange information of common interest.

Soviet scientists are even isolated from each other. They seldom change jobs. National scientific conferences are few in number, visits to other institutions even in the same city let alone in other parts of the country are infrequent, and multidisciplinary projects within institutes are the exception. Indeed, most Soviet scientists have historically confined their professional interests to the narrow problems for which they are responsible. Over the years they have learned, as has the general population, that too much knowledge concerning the activities of others can often create unfounded suspicions about their interests and motivations.

The gloomy picture of Soviet science painted by Sagdeyev has been reinforced by personal observations of many Western visitors to the USSR. Over the years I have walked through dozens of Soviet research facilities, toured many industrial plants, and discussed developments with hundreds of Soviet scientists and engineers. While I am always discouraged by the cramped space, idle equipment, and narrowness of Soviet perspectives, I am repeatedly amazed by the Soviet ingenuity in working around physical shortcomings. From arranging computers in parallel to support the space program to cutting gaskets from discarded tin cans for repairing field equipment, Soviet specialists have learned to use what they have to the fullest extent possible. Though the environment in the USSR hardly seems conducive for high-quality research activities, many Soviet scientists regularly obtain results of broad international interest in a variety of disciplines. No matter the hardships, Soviet science will survive and surprise.

*   *   *

Since the beginning of the Communist era, Soviet engineers have been called upon to play a key role in developing the economic base of a society whose goal is to emulate and eventually to surpass the West in its technical achievements. Steel, concrete, electricity, machinery, and chemicals comprise the tools of development. These are the domain of the engineers. Engineers are to guide the industrial workers, and the workers are to be the vanguard of the modernization effort based on Socialist principles, principles which are currently being redefined.

Engineers have been well suited to the task of managing enterprises—Soviet style. Directing production has been considered to be a technical task: the managers were given quotas to meet and were not bothered with determining costs of raw materials or prices of their products. Nonetheless, they had to be ingenious in compensating for unreliable supplies, unsuitable production equipment, and inadequately trained workers. They relied on the party and the trade union to ensure that the workers received appropriate treatment and that the benefits of successful production efforts were distributed in accordance with Communist principles.

While many enterprise directors have historically been engineers, there is a growing trend to recruit directors from other professions as well who are sensitive to economic concerns and to the intricacies of the overall planning and allocation system. Still, an encounter with a chief engineer at a heavy industrial plant in the USSR quickly dispels doubts about the continuing influence of the engineering profession on production.

The chief engineer remains in charge of activities on the factory floor. He commands the attention—and usually the respect—of both enterprise management and the workers, like a drill sergeant ensuring that schedules are maintained and that problems are resolved at the lowest possible level. On a number of visits to Soviet factories, I have witnessed directors take back seats to the chief engineers during discussions of specific production activities. Obviously there are strong directors who do dominate, but the chief engineer will continue to have high status in the USSR. His status does not seem much different from the position of counterparts in Western plants, but it is striking that the Soviet chief engineer has retained such authority within an autocratical system where directors of organizations have been the dominant figures.

In many ways the engineers have provided the wherewithal for transforming a backward Soviet society into a modern industrial state. The USSR now leads the world in timber, oil, gas, and steel production. It ranks second after the United States in electricity, second after China in cement, and third after the United States and China in coal.[7] Soviet ships ply the seas of the world, and Soviet trains and planes continuously span a vast land mass. But the Soviet approach has only partially succeeded. The quality of Soviet products is generally poor with few takers on the international markets and complaints and disenchantment at home. The concept of quality control has received little more than lip service except in the military sector. Even in Eastern Europe, ordinary citizens constantly curse the quality of Soviet products which are imported as the result of compulsory trade arrangements. This reality was brought home to me while speeding along the highway from Sofia to Plovdiv in Bulgaria in 1985. We experienced a near fatal calamity as the tread separated from the rest of a tire on our newly imported Soviet car.

Modernization of Soviet industrial facilities is finally in vogue. It had always seemed simpler, and certainly much more photogenic, to build new plants or to install new production lines rather than to upgrade existing facilities in a systematic and economical manner. Now hundreds of outmoded and inefficient production processes clogging the country must continue to operate lest the entire economy break down. On the other hand, punctuating the countryside are a few outstanding factories employing the latest world technologies. In most cases these are based on imported foreign technologies. This foreign technological influence has clearly strengthened Soviet capabilities in specific areas, but it has had a relatively minor impact on the overall performance of the economy.

At some of the new facilities the Soviet workforce has great difficulty coping with advanced technologies. Both the engineers and the technicians are simply out of date with new technological developments and particularly with those that involve electronic components. Mechanical systems have been installed by either foreign or Soviet specialists who promptly depart after installation. The workforce, including the engineers, is often uncertain about the intricacies of the new controls and frequently uses the old manual systems while the automated equipment stands idly by. In some cases automatic control systems become inoperative because of the lack of key spare parts or the absence of repair skills at the enterprise.

Soviet engineers cope with the lack of specialized materials and components and frequently rely on older technologies to design and manufacture modern products. This approach has led to a propensity to overdesign by adding bulk and weight to ensure performance in products such as automobiles and airplanes, which operate quite well but inefficiently by Western standards. In some cases, the engineers have no choice but to design products with short lifetimes or with requirements for frequent maintenance checks. Many advanced alloys, lubricants, electronics, and other high technologies which buoy industrial processes and products in the West simply are not readily available in the USSR, where older technologies dictate other approaches as the only alternative.

Traditionally, Soviet planners have ignored the real cost of energy, and particularly energy used throughout the industrial complex. Only when international oil prices fell in the early 1980s, which sharply reduced earnings from oil exports, was serious attention directed to the implicit policy of subsidizing energy for almost all applications in the USSR. The opportunities for energy conservation within industry appear substantial, provided there are internal price adjustments to force the ministries and enterprises to adjust industrial processes accordingly.

The current waste of energy clearly needs attention, particularly as the ambitious Soviet nuclear power program slows down due to safety concerns. Many relatively simple engineering adjustments could save large quantities of energy—improved design and construction of buildings, better operating procedures in power plants, and better organization of industrial processes. Without economic incentives, however, the likelihood of such adjustments on a broad scale seems low.

The industrial workforce in the USSR has lost much of its earlier zest which was entwined with patriotic loyalty to the workers' state. As in other countries throughout the world, the workers have become interested primarily in their own well-being in the short term—the hours they work, the energy they expend, and the pay they receive. Respected Soviet social scientists documented this development concerning the laziness of Soviet workers in unpublished manuscripts that circulated quietly throughout the USSR and the Eastern bloc during the mid-1980s. They underscored the weaknesses of an economic system devoid of real incentives. I first heard about these manuscripts while visiting Eastern Europe in 1985, and these stories were subsequently confirmed during my trips to Moscow the next year. Now, discussions of low worker productivity are on the agendas at many government and party meetings in the USSR. This is one problem that engineers cannot solve.

Finally, under the headline "Decline in Prestige of Engineers," a Moscow journal recently noted that young engineers at research institutes spend 60 percent of their time at menial tasks. They become construction workers, organizers of sports competitions, and members of "people patrols." They work as couriers, typists, and accountants, hardly tasks that inspire engineering creativity or hone basic engineering skills. Thus, the frequently quoted statistic showing that one-half the world's engineers work in the USSR is very misleading.[8]

*   *   *

Military science and engineering are sharply segregated from civilian activities in the USSR. In the view of the noted Soviet historian Roy Medvedev, Americans conclude "Since Russia can't run a hotel, it can't build a rocket either." He adds, "They don't realize that we put everything into rocketry."[9]

The dichotomy in the USSR between world-class achievements in the military sector and the problems of quantity and quality in the civilian sector is truly a startling phenomenon witnessed nowhere else in the world. Will the Soviet Union be able to sustain over the long term a military capability which rests on the leading edge of technology within a country whose aging industrial base is falling further and further behind international technological standards?

The explanation for the gap between military and civilian achievements lies in priorities—priorities at all levels. The highest quality design and production facilities are dedicated to military production. Raw materials and supplies needed to keep these facilities functioning are skimmed off the top during the national planning process and at the operational level when planning targets and delivery schedules are in jeopardy. Consequently, many civilian enterprises competing for the same materials and supplies inevitably come up short or must be content with inferior materials and supplies.

Many of the best scientists and engineers work in the defense sector, often recruited while still students at higher educational institutions. They are provided with at least indirect access to Western technologies obtained through a variety of open and covert mechanisms. They are generously rewarded financially for particularly noteworthy accomplishments. In sharp contrast to the civilian sector where the customer has little influence on manufacturing practices, quality control and efficiency have long been concepts incorporated into the consciousness and actions of personnel throughout the industrial complex which serves the military. The military establishment is a customer with rigorous acceptance standards and considerable clout to enforce these standards.

This system of priorities for military projects has similarities to the wartime effort in the United States in the 1940s when we diverted our best resources and people into supporting defense programs. We had only minor difficulties in sustaining this effort for a few years. However, in contrast to the Soviet scene, not only could we see the light at the end of the tunnel early in the effort, but we also had a tremendous stock of high quality civilian products to see us through the years of shortages and substitutes.

The Soviets are having increasing difficulty sustaining a comparable "wartime" effort over many decades. There is a restlessness among the general populace over the retarded state of the civilian economy, a restlessness that is beginning to stir in a more visible manner during the glasnost era. Also, a completely closed military technology effort denies the military sector important benefits from direct interactions with worldwide technological developments. The international technology collection efforts of the military intelligence services provide only limited compensation in this regard.

The Pentagon's annual publication *Soviet Military Power: An Assessment of the Threat, 1988* highlights technology as the key to military capabilities of the two superpowers. This publication wisely notes, however,

> Competition between the Soviet Union and the United States is not purely technological . . . technology does not, in and of itself, revise any of the military balance . . . it is how well technology is applied, and how thoroughly its contributions to military operations are absorbed by those who use that

technology, that have the greatest impact upon the military balance.[10]

Thus, as military operations become more diverse and as military systems become more complex, the country with the broadest base of technology and the highest level of technological literacy among the population has a tremendous advantage in the military competition. This message has been highlighted during the debates over SDI. As we have seen, in the Soviet view the question is not whether SDI will destroy a small number or a large number of incoming missiles or whether SDI will ever be deployed. The real issue is whether, under the SDI umbrella, the United States will develop a vast array of new electronic and related technologies which can be oriented to many types of military applications. The Soviets are clearly concerned that they cannot compete successfully in this arena, for they do not have the mature industrial base and supporting technological infrastructure that is available in the United States to support such a program. This concern over the across-the-board weakness of Soviet industry is at the core of the support, albeit tenuous, of the Soviet military establishment for perestroika.

Several hundred Soviet R & D institutions devote a substantial amount of their effort to support defense programs. Most of these institutions are affiliated with the ministries and enterprises responsible for production; and a number of them also contribute to civilian R & D in fields such as aviation, shipbuilding, and electronics. A few of the institutions are directly subordinate to the Ministry of Defense. These institutions are concerned primarily with defining military requirements for weapons systems and improving effective utilization of new military systems. Many institutes of the Academy of Sciences and individuals and groups of scientists at higher educational institutions receive funds to work on research projects of interest to the defense establishment.

There are several similarities between the Soviet and American approaches to military research. Key defense laboratories in both countries sharply orient their R & D programs toward military requirements. Both countries rely heavily on industrial facilities to design and to develop new systems and system components. Also, they rely increasingly on the academic communities to investigate scientific areas of future interest.

But the differences are very pronounced. For example, American aerospace and other high-technology firms compete against each other for military contracts, and during this competition they continually advance the state of the art. Competition of this sort is less common in the USSR. Only in the field of aeronautics do Soviet design bureaus concerned with missiles and aircraft in effect compete for Soviet "contracts." Second, laboratories of our Department of Defense have long had technology spin-off programs to help the civilian sector—in construction technology, in electronics, in advanced materials. This concept is in its earliest discussion stages in the USSR.

Perhaps most important, American scientists working on defense projects are reasonably well integrated into the broader American scientific community, participating in scientific conferences, publishing unclassified papers, and consulting regularly with a wide range of colleagues. Mobility of American scientists between jobs located in different institutional settings has become perhaps the most important mechanism for technology diffusion, a mechanism virtually unknown to scientists, and particularly military scientists, in the USSR.

Looking more closely at Soviet military technology, we see many technological advances that have been incorporated into very potent weapons systems with capabilities equal to or better than our systems. Some of these technologies had their origins in the West while others are clearly Soviet designs. Many of the most impressive Soviet technological breakthroughs are regularly cited in Pentagon assessments, and particularly assessments set forth to justify budget requests. For example, Soviet submarines are noted for their speed and diving ability, in part because of light, strong titanium hulls. Modified Soviet MiGs hold world speed and altitude records. Soviet achievements in rocketry fill a variety of record books. With regard to conventional armaments, Soviet armored vehicles have several advanced offensive and defensive features.

Even with these impressive accomplishments, limitations abound. For instance, Soviet submarines are sometimes unreliable, partially due to problems with their nuclear reactor propulsion systems. Soviet aircraft also have problems of reliability. Some developing countries are reluctant to use Soviet commercial aircraft, and Soviet fighter planes which have fallen into Western hands have displayed outdated electronic systems and very short-range fighting capabilities. The vaunted space program has its limitations. Again, reliability is a key concern, with the lifetimes of Soviet satellites much shorter than the lifetimes of comparable American satellites, probably due to limited propulsion and control capabilities.

Many relatively crude Soviet technologies generally perform as designed, and weaponry based on the technologies of the 1960s and 1970s remains a potent factor. Indeed, in many military settings American systems may be much too sophisticated and costly. Advanced technology is not always synonymous with fighting capability, although in many situations a direct correlation exists.

*   *   *

As military systems continue to grow in sophistication and complexity and to involve innovations of every kind, the overlap between military and civilian technology in both countries becomes increasingly apparent. The defense establishment in the United States has historically drawn heavily on civilian technologies for materials and fuels for its weaponry and for components of its fire control, navigation, and detection systems. In contrast, a great gap exists in the USSR between defense needs and the relatively low level of available civilian technologies. Many important ceramics, chemicals, and electronic components, for example, simply have not been developed for civilian uses. One objective of perestroika is to close the military/civilian technology gap by increasing the flow of technical achievements from the military to the civilian sector and eventually from the civilian to the military sector.

A brief look at Soviet capabilities in the fields of computers, machine tools, and advanced materials will help set the stage for consideration of US policies which must address technologies with important applications in both the civilian and military sectors. These three fields have long been focal points for debates over our trade and technology transfer policies.[11]

Intensive use of computers began in the United States in the mid–1950s. A comparable effort in the USSR had its origins about 10 years later. In both countries, the first 10 years of development saw a sharp growth in the production and use of mainframe computers. The growth continued rapidly in the United States with high speed mini- and microcomputers appearing in the 1970s. By the mid-1980s the number of powerful computer systems totaled in the millions. In the USSR the growth has flattened in recent years, and the powerful computers available within the country number only in the hundreds of thousands. The 10-year lag has stretched to 15 years or more, with little sign of early closing of the gap.

A particularly telling indication of the poor state of Soviet computers is a statement repeated to me several times by Bulgarian specialists and quickly confirmed by Soviet specialists: the computer power per capita in Bulgaria in 1987 added up to three times that in the USSR. Soviet specialists can tolerate lagging behind the United States, Japan, or Western Europe. But to be compared with Bulgaria? Tolerance does have its limits.

The Soviets can of course mobilize large quantities of talent to design computers with respectable speeds, although not rivaling the speeds which had been attainable in the West, say, five years earlier. However, transferring technology from development to production to use takes several times longer than in the United States. The Soviets are skilled at producing the first several prototypes of sophisticated equipment, often at an experimental factory associated with a research facility. Moving to mass production, on the other hand, is plagued with problems of quality control. The compartmentalized nature of Soviet activities greatly hampers effective installation, use, and maintenance of computers. And once the computer is delivered to the doorstep of the users, they are immediately on their own.

Visits to Soviet scientific institutions, large business establishments, or industrial enterprises usually reveal startling arrays of computers of all sizes, capabilities, ages, and origins. The Novosibirsk Computer Center, for example, relies on a Soviet mainframe which I first saw there in the mid-1960s, on a newly acquired Soviet mainframe, on a more powerful minicomputer imported from the United States in 1980, on recently acquired personal computers from the West and from Asia, and on an assortment of computer components and systems obtained from several East European countries. An enormous amount of effort is devoted to ensure the proper functioning of these systems, to develop a limited degree of compatibility among the different software operating systems that they employ, and to train specialists who can use them efficiently.

Each time I am in Moscow, I visit the single retail store where personal computers may be purchased. During each visit I receive the same reply to my query as to the procedure for purchasing a personal computer. "We have no computers in stock right now, but we are expecting a shipment in several months. Come in at that time, and we will take your order." While Soviet computers are slowly becoming available to both institutions and individuals, the only effective approach to gaining possession of a Soviet computer is to work through an acquaintance who is employed somewhere in the chain between the factory and the retail store. Also, it is easy to understand why Soviet visitors to the United States try their best to save money to purchase a computer which can be taken back to the USSR.

As to computer software, worldwide developments since 1980 have been revolutionary. Much of the effort in the West has been devoted to computer applications for word processing, publishing, and printing; to storing, formatting, and accessing large amounts of data; and to supporting systems for financial, medical, and other everyday needs. A much greater portion of the modest Soviet effort has been devoted to automation of industrial production processes.

Many well-qualified Soviet specialists work in developing software. Their rigorous training in mathematics provides a strong basis for responding to computer operating and programming requirements. The software explosion in the West will eventually impact the East as well and will assist these Soviet specialists in areas which have eluded them in the past.

With regard to machine tools for cutting and shaping metals, development and production of many types for use throughout industry have long been a high priority within the USSR. When the United States expanded embargoes on the export of advanced machine tools in the mid-1970s, the Soviets decided to devote the necessary financial and manpower resources to the development of their own world-class machine tool industry rather than to rely on uncertain Western sources of equipment.

In the last 10 years they have succeeded in producing a few machine tools which now compete on the international market. Several Soviet plants are producing large quantities of these tools which are in demand at home and abroad.

The Soviets face a great handicap in developing and producing the microprocessors and related components for the control of modern machine tools, given the primitive state of their electronics industry. In some cases, they export the machine tools without the electronic components. For example, Canadian importers buy stripped-down machine tools from the USSR and then install Japanese electronics in Canada. In other cases, and particularly with regard to relatively simple machine tools, the Soviets have developed very clever software which permits the use of their own bulky systems for controlling the operations.

The East European countries play an important role in providing the USSR with machine tools and machine tool components for industrial use. Soviet production levels cannot yet satisfy local demands as well as export requirements. East Germany, Czechoslovakia, and Bulgaria are providing systems and components for the large Soviet market, and Hungary has gained a reputation as a world leader in software systems for automated machine tools.

One significant area of Soviet excellence, as we have noted, is titanium metallurgy. Lightweight titanium alloys have very desirable heat-resistant properties at high temperatures. Advances in this field are particularly important in the manufacture of modern submarines and aircraft as well as turbines for power plants. During my visit to the Leningrad Metal Works in December 1988, the deputy director beamed while discussing the properties of the titanium-tipped turbine blades produced in the factory. The plant recently joined forces with an Italian engineering firm in bidding for a major contract to supply power equipment to Greece, including turbines manufactured in Leningrad.

The USSR needs a substantial overhaul of many of its production facilities as a step toward larger production of high-quality steels. In particular, many of the antiquated open-hearth furnaces should be replaced. A major shifting of priorities, technologies, and scientific talent is required if the USSR is to compete effectively in the West.

In the 1950s and 1960s, when the training of Soviet chemical engineers was at its peak, Soviet research was at a world level in investigating lightweight ceramics which could be used in both civilian and military products subjected to high temperatures. Now the USSR has slipped badly as the number of outstanding young specialists entering the field has dropped. Many specialists in the overall pool are narrowly trained and not ready to shift into new and promising areas. For similar reasons, the Soviets continue to lag in the field of chemical polymers with many applications in both the military and civilian sectors.

The science of advanced materials in the USSR has suffered from a preoccupation with basic research to the detriment of sufficient emphasis on applications. And once again, the common problems of inadequate experimental instrumentation and bureaucratic insulation between the researchers and the production organizations have magnified the dilemma.

Exemplifying this insulation, *The Economist* writes,

> If you want to see the best Soviet technology, go to Japan. . . .
> The Japanese are superb at spotting a Soviet idea (for example,
> for making continuously cast steel), buying the license and
> racing ahead to apply it across the industry; back in Russia,
> meanwhile, the technology spreads slowly, if at all. Russia's
> problem is not any shortage of inventiveness. It is lack of
> incentives to make the jump from research to the production
> and diffusion of a new process . . . the technology had spread
> to only 12 percent of the Soviet steel industry 27 years after its
> first use at the Novolipetsk metallurgical factory; in West Ger-
> many it had spread to 62 percent, and in Japan to 79 percent.[12]

*   *   *

Capabilities in science and engineering spring from a nation's educational system. Now, as high technologies enter the scene, a wide diversity of well-developed scientific skills throughout the Soviet economy is more important than ever before. Only through a reinvigorated educational system will the Soviet Union develop the capability to apply science and technology to economic requirements on a broad basis. However, decrying the failures of the past, the Soviet Minister of Higher Education in 1987 described the approach to education in the USSR as follows: "In our pursuit of technological achievement, we tried to cram into our school children and students an amount of knowledge they could not possibly digest."[13]

There are about 900 institutions of higher education in the USSR. Education ministries in the 15 republics direct most of these institutions. A few of the most prestigious ones are administered directly by the national government in Moscow. This organizational fragmentation has often been cited as a problem in effecting across-the-board adjustments and improvements in the Soviet approach to education. In contrast, the United States boasts of the diversity of its more than 3,000 institutions of higher education. Most of our institutions have independent Boards of Trustees; diversity rather than standardization is considered an important strength of the US approach.

The Soviet higher educational system has historically emphasized technical training. The output of graduates in science and engineering each year greatly exceeds the output of graduates in the United States. The Soviets say they produce four times the number of engineers and twice the number of scientists that graduate in the United States. The definitions of scientists and engineers are not comparable in the two countries, and these comparisons may not be entirely fair. For example, many types of social scientists and technicians are included in the Soviet definitions of scientists and engineers. Still, by any measure, the Soviet technical training effort is huge.

An important aspect of the Soviet approach is the heavy emphasis on evening and extension courses. Forty percent of the students are enrolled in these programs, including many which are held after working hours at enterprises and other convenient locations. Some educators within the USSR have reservations concerning the quality of this type of technical preparation for such a large segment of the scientific and engineering community. They are not convinced that educations received on a part-time basis can substitute on a very large scale for full-time educational experiences.

A few of the Soviet science and engineering faculties are of very high quality. These are concentrated in Moscow, Leningrad, Novosibirsk, Kiev, and several other major cities. Hundreds of other science and engineering faculties stretch from Lvov on the Polish border to Vladivostok. Many are less impressive—in terms of the quality of the students, the preparation of the instructors, and the availability of facilities and equipment.

At the undergraduate level, students attend many lectures during the week and on Saturdays. They rely heavily on lecture notes and on a limited number of textbooks. The students simply do not have time for supplementary reading or independent study due to the number of required class hours. Examinations are often oral. Each student must complete an independent project during the final year, which is usually the fifth year of study. A few science students spend their final two years affiliated with research institutes—sometimes institutes of the Academy of Sciences—where they are exposed to advanced research and where they often initiate their job searches.

Almost all educational institutions suffer from inadequate research capabilities of their own. The Soviet philosophy has historically been to concentrate research in the institutes of the Academy of Sciences and in institutes of the industrial, agricultural, and health ministries. This approach is quite different from the Western approach which emphasizes the mutual reinforcement of education and research located under the same roof.

Under new educational guidelines, Soviet science students are to be provided greater opportunities to participate in research. In practice this philosophy works well for the best students, who pair off with the most important professors. However, the average student has limited exposure to significant research. While 70 Soviet educational institutions have recently been selected as centers for an expanded research emphasis, including construction of better research facilities within the institutions, the opportunities for research even within these institutions continue to be limited and very uneven for faculty and students.

The situation with engineers, and particularly the weaker engineering students, is better, in view of the ample opportunities for practical work at enterprises in connection with their studies. Still, the lack of emphasis on engineering research within the educational programs reduces the likelihood that latent creativity will be nurtured and that innovation will become a significant concern of the graduates.

With regard to advanced degrees, the best science students often go directly from the universities to research institutes. Some institutes have the authority for granting advanced degrees when the young scientists continue their education while working on research projects of interest to the institutes. Another common route is for an aspiring young scientist to be assigned by his or her employer, such as an enterprise or industrial institute, for full-time study at a leading university or Academy institute. After two, three, or four years, the successful scientist earns a Candidate degree, which is roughly equivalent to a PhD. The scientist then returns to the original place of employment to continue his or her career. There are many variations on this approach, including encouragement of older scientists to take time off to earn their Candidate degrees rather late in their careers.

An advanced degree requires a thesis, as in the United States. The student is usually required to demonstrate the usefulness to society of the research reported in the dissertation. The student is also expected to publish articles on the thesis topic. Prior to acceptance, the thesis is commonly circulated to institutions and individuals who may be interested in commenting on it. The student then defends the thesis in a forum open to the public which has been advertised in advance. I have attended some of these sessions; sometimes they are filled with scientific controversy, while on other occasions they seem almost *pro forma* in satisfying an academic requirement.

The final degree in the USSR is the Doctorate, which has traditionally been reserved for a limited number of senior scientists who have prepared numerous scientific publications. Late in their careers, particularly successful senior scientists take off a year or more to compile their most important works into a dissertation which qualifies for this prestigious recognition. Recently, in efforts to lower the average age of the recipients of this degree, the education authorities have developed procedures for scientists under 40 to receive sabbatical leaves of up to three years to work on a doctoral thesis.

As in the United States, discontent in the USSR is growing over the quality of science education at the secondary school level. National achievement levels seem to be slipping, even in mathematics, a field of traditional strength. Of special concern is the stifling effect of reliance on rote learning within curricula which are too ambitious for available class time.

A few special Soviet secondary schools have received worldwide recognition for the achievements of their students in mathematics and science. Also, the Soviet annual "olympics" for the most outstanding students in these fields is well known, and the results of the participants are impressive. However, as in all countries, the achievements of a few do not reflect general capabilities.

Overall, the Soviet educational system gives great attention to rigorous training in the basic scientific disciplines with a heavy emphasis on technical subjects at all levels. While there has been some decline in the interest in science among the Soviet youth, the competition for admission to the best universities and engineering schools remains intense. The graduates of these institutions are highly skilled in the fundamentals of science and engineering. However, the educational experiences even for the best and the brightest are often narrowly based. Individually tailored educational programs and interdisciplinary programs are seldom available. This tradition of narrow educational preparation is undoubtedly a contributing factor to the general lack of personal incentive for mobility of specialists between jobs later in life.

*   *   *

Scientists and engineers in every country are dissatisfied with the level of financial support for their activities, and the Soviet Union is no exception. They are impatient with the slowness of personal promotion opportunities and they crave greater peer recognition. They bemoan the lack of appreciation within the government and industry, and indeed within the general population, of the importance to society of the results of their research efforts. Can incentives and investments improve the opportunities for better science and more rapid industrial growth in the USSR?

Perhaps the most stifling aspect of research in the USSR is the attitude of resignation that often develops among researchers, and particularly among the researchers who do not have academicians as mentors and who do not find themselves in the select laboratories of the country. While the standards in their laboratories may be high and the research activities important, they have little hope of moving from their individual laboratories or of stretching beyond the limits of the laboratories' traditional research profiles during their entire careers. There are few opportunities for peer recognition outside their immediate research groups. In a short time, research becomes simply a job, a comfortable job offering a reasonably good salary and considerable prestige, but a job largely devoid of the excitement that initially attracts many bright students to science in all countries.

Still, in most areas Soviet science is moving forward. Technology moves more slowly, but technological breakthroughs are well ahead of the application and diffusion of these developments. Constrained by a tradition of not sharing important technological advances, Soviet industry has great difficulty absorbing any new development on a wide scale. Individual plants rather than industrial sectors are the principal beneficiaries of technologies acquired abroad or developed internally. New economic incentives for using and diffusing technologies which will enhance productivity and quality are repeatedly cited as the linchpin to industrial progress in the USSR.

Everyone favors incentives for innovation, particularly budget officials, who prefer incentives to more expensive approaches to promote economic growth. But what is meant by incentives? Monetary rewards for harder work, for closer attention to the needs and interests of customers, or for better organization and management of research or manufacturing activities? The Soviets talk about all these types of rewards.

The enthusiasm for incentives rests on two premises. First, not only is there room for improvements, but the overall system can accommodate and reward approaches which will lead to improvements. Second, improvements are technically feasible and can be reflected in research or manufacturing. As we have seen, these conditions are not always present in the USSR.

Further adding to the difficulties of fostering more productive research and engineering, individual firms and research institutes may be accustomed to making all components of a product or conducting all aspects of a research project. Reliance on other Soviet institutions whose dependability is in question is to be avoided even though there may be comparative advantages of specialization within different organizations. In contrast, the opportunity to draw on a variety of organizations with many types of skills has proven to be very important in improving efficiency and productivity in the West.

Innovative ideas become reality only if they are tied to investments. Up-front financing is usually necessary to initiate a new type of production activity. While the Soviets have attempted for many years to link their investment plans and their technological development plans, the results to date have not been very encouraging. For instance, the key industrial areas which are being highlighted by Gorbachev—machinery, chemicals, and electric power—have been priority investment areas for more than 25 years, yet technologies have lagged. Given the severe limitation on investment resources in the USSR even for priority areas, the likelihood of major new investment thrusts to stimulate the use of more productive technologies does not seem high.

Of critical importance, support from the Soviet military leaders for a reduced emphasis on military spending cannot be taken for granted. They may reluctantly accept the trade-off of less funds for procurement of military goods in exchange for greater investments in a broad and more sophisticated technological and industrial base. They will probably be more than willing to share the burden for development of technologies with both civilian and military applications. However, they are clearly wary about eroding the favored status of the military sector. They will surely become impatient with delays in implementing civilian R & D programs. Finally, they will become increasingly apprehensive if the United States continues to widen the military technology gap and uses this superiority for political advantage.

Just as the strengths and weaknesses of Soviet technology are often difficult to explain, the effects of imports of Western technologies cannot be easily predicted. Sometimes acquisitions of technology from the West reinforce Soviet efforts and accelerate the development of an indigenous technological capability. In other cases, Soviet dependence on Western technology has led to an atrophy of emerging Soviet R & D, with the result being a greater rather than a smaller technological gap.

The Soviets are fearful of long-term dependence on the West in any area, but they are increasingly faced with this reality. Technology transfer is a people problem. Until Soviet specialists become more fully attuned to the international technological movement, the USSR has little chance to capitalize on Western technological achievements short of simply buying Western hardware and management skills.[14]

CHAPTER 4

Changing the Course through Perestroika

Anything that won't sell, I don't want to invent.
Thomas Edison

Gorbachev has set the goal of Soviet achievement of world technological standards in all areas of the Soviet economy—automated factories, highly efficient power plants, modern highways across the countryside, plentiful high-quality consumer goods. He talks about reaching this goal in many branches of industry during the 1990s. His advisers predict that within 25 years the USSR will attain a level of economic productivity and a standard of living that are second to none. But bear in mind, these targets are to be realized by a country where 60 percent of the workforce is currently engaged in manual labor; where more than 50 million pensioners depend on heavily subsidized goods and services; and where the economy has been labeled by Soviet economists as "no-growth" for more than a decade.

For many years the Soviet leadership has promised to overtake the industrial accomplishments of the West, and these new assertions continue a favorite form of Soviet fantasizing about achieving the impossible. However, the long-term time frame now being espoused is still a refreshing contrast to the more bombastic predictions of the past. Gorbachev's sincerity is probably genuine, and his optimism spurs the nation forward. Nevertheless, the race with the West was lost before he was summoned to the starting line, and now his task is to stay as close to the leaders as possible until he passes the baton to his successor.

With regard to the short term, current plans call for financial and technical manpower priorities in industrial development: machine tools, computers, electronics, steel, and chemicals. Soviet leaders believe that an overhaul of the entire industrial base is an essential first step before the benefits of perestroika are felt throughout the economy.

Still, the workers are impatient; they crave immediate benefits. In recognition of the need for prompt and visible improvements in the general standard of living, Gorbachev has singled out three sectors for particularly urgent attention, namely, food processing, housing construction, and health services. The government has promised substantial new investments in these areas; and these are Soviet priority areas for international joint ventures.

In the meantime, the Soviet political elite is reluctantly giving up some of the personal privileges that have placed them outside the daily hassles of survival in the USSR. For example, the Moscow press recently revealed that the state-subsidized grocery list for the weekend summer homes of high officials in the Ryazan region southeast of Moscow for the first six months of 1988 included 394 kilograms of caviar; six tons of crab, pâté, and other delicacies; and 565 kilograms of cured sturgeon.[1] Now these aficionados of high living are beginning to lose their access to special stores, imported goods, and many on-call delivery and repair services. According to Gorbachev's philosophy, they too must learn to cope with long lines at every turn, with housing shortages that throw newly married couples and parents together for many years and thereby contribute to the growing divorce rate, and with an unquenched thirst for Western luxuries which they can see only in Western films or during travels abroad.

*   *   *

Agricultural deficiencies loom large and are now one of the targets of Gorbachev's efforts. Throughout his recent career, Gorbachev has been close to agriculture as the party chief in the Stavropol agricultural region and as Brezhnev's top aide for agricultural development. Yet, somewhat surprisingly, most of the initial attention in the USSR during the 1980s for more effective use of science through economic reforms has been directed to the complicated industrial sector.

In contrast, early reforms in China were heavily concentrated in the agricultural sector: food was the first priority before raising the calls for harder work in the factories. Soviet economists belittle any comparison between agriculture in China and in the USSR, agriculture which they emphasize evolved from entirely different historical and cultural roots. Still, the Soviet leaders somewhat belatedly recognize the urgent need to restock the markets at affordable prices if their repeated calls for reform throughout the country are to be supported by the population. Greater personal incentives for the workers and the peasants are to be the key in the agricultural sector. Gorbachev wants farmers to be "masters of the land" again. He has little patience with the Russian adage, "We pretend to work; and they pretend to pay us."

Apologists for the perennial agricultural shortfalls in the USSR have usually blamed the uncertainties of the weather. Weather is but a part of the story. Poorly developed road networks quickly turn to mud quagmires in the rural areas. Storage facilities are inadequate to prevent the rotting of crops following harvest. Mismatches between the distribution of agricultural equipment and local needs for mechanization are commonplace. Technological fixes to address these and other obvious problems are critical. As an immediate step, the government has promised expenditures of 40 billion rubles to improve roads in the agricultural heartland. However, better hybrids, massive investments in tractors, intensive use of agricultural fertilizers, and other technological approaches have not in and of themselves provided promised returns in the past. Thus, the uninspired fortitude of the farmers in the past and at present is also a major concern.

During the late 1920s and 1930s, Stalin planned to solve the problem of lagging agricultural production through forced collectivization of the peasants. In the process millions died from starvation. The collective farms which call for common ownership of the land and common sharing of the harvest have varied widely in their productivity over the years, and in some cases they have been very successful. However, over time they frequently have become more like state enterprises which simply pay each worker a fixed salary, with financial incentives for the individual farmers tied to overall productivity of the collectives gradually fading in significance.

Within Soviet planning circles, the economic aspect of agriculture has begun to drown out the remnants of the ideological emphasis that depersonalized the farmers and hampered past efforts. The Soviets now recognize the importance of a small but dynamic private sector. A significant step in this regard is the new policy of offering individual family farmers 50-year leases for plots of state lands and rental agreements for use of state equipment. Relaxation of some of the price controls on agricultural products of both the state and private sectors is also needed; the resulting higher prices may be painful to the consumers in the short term but should be helpful in raising production levels and bringing down prices in the long term. Also, improved scientific management of the farmlands must become more of a reality than a slogan if production is to rise.[2]

*   *   *

Turning to the problems of upgrading the performance of Soviet industry, American skeptics doubt that the Western concepts of marketing and profit motivation can be transferred to an economy which must rely on a heavy element of state ownership and central planning for the indefinite future. Nevertheless, the centerpiece of the Soviet economic reforms is the designation of each state-owned enterprise as a self-sustaining profit center with its own responsibility to ensure that it operates in the black. Some sales of most enterprises, say 50 percent, will be guaranteed under contracts with the State ministries. These ministries are to decide which production is absolutely essential for the continued functioning of the economy and then control all aspects of the production and distribution through contracts with the appropriate enterprises. The remainder of the sales are to be carried out through a free market system which is only now being defined and developed. Contrary to the spirit of perestroika, some enterprises are reluctant to enter into the uncertainties of the free market and are trying to maintain state contracts for 80 to 100 percent of their sales. "Let others experiment with the new system while we are assured of at least a steady income level," they say.

Industrial modernization in the USSR requires extensive renovation of many, if not most, of the production facilities, since the costs of simply scrapping obsolete facilities in favor of new modern factories on a wide scale are prohibitive. Indeed, the traditional bias in the USSR toward building new facilities rather than retrofitting existing plants must now be tempered with solid economic analysis of the trade-offs between new and old. Renovation in turn will be difficult, for it requires extensive custom building of machinery which will function properly within existing facilities. Unfortunately, manufacture of customized machinery is not a strong suit in the USSR: the system has rewarded managers who are successful in serial production of identical machines, because production targets have been based on quantity. In addition, production lines must stop when new customized machinery is being installed in existing facilities. The managers and workers ask, "Will the new equipment really outperform existing lines and warrant downtime for installing the equipment—downtime which can jeopardize fulfillment of current production quotas?" But in many plants there will be no alternative to major revamping of the existing lines.

Better quality control is essential. However, managers lose interest in quality when, in view of nationwide shortages, they are able to sell everything they manufacture regardless of quality. Thus, efforts to promote the idea of quality control find little receptivity in the civilian sector.

Regardless of changes in the incentive system, neither enterprise managers nor their workers will see the promised bonuses, improved living conditions, and heightened personal status unless the technical tools at their disposal are adequate to the task of raising worker productivity. Eliminating the hand cranks and the shovels, mating new equipment with existing tools and apparatus of another vintage, installing automatic control procedures, and operating quality control systems are not easy tasks. They require both technical know-how and investment capital. Technical know-how is the challenge of the educational system, the management system, and the scientific research and development system.

*   *   *

Repeated references to Soviet military achievements must color any discussion of reforms in Soviet industry. Surely, if Soviet space vehicles can reach Mars and their nuclear weapons can destroy Los Angeles, Soviet engineers can design buildings that do not begin to crumble as soon as the first occupants arrive, and Soviet workers can produce jeans that do not rip during the first stretch. It is not surprising that the Soviet leaders give high priority to enlisting Soviet military capabilities to bolster the civilian industrial base on a broad scale. In the past some Soviet plants have produced both civilian and military products—trucks, planes, ships, cranes, electronic equipment; but the military products have always received priority over civilian requirements. Now a number of highly successful managers of Soviet military production activities have been moved to new posts where they have responsibility for the management of many more civilian production activities at the national level. In some cases, reorganizations have placed both military and civilian plants completely under their control.

Representatives of the State Planning Committee repeatedly underscored the importance of drawing on Soviet military experiences in boosting civilian production during a seminar I attended in Moscow at the end of 1988. Reflecting recent Soviet political statements calling for every military plant to produce something for the civilian economy, a top official of the committee was particularly interested in the conversion of military production lines to civilian lines. As an example, he wanted to know how to produce dairy equipment side by side with equipment to support missile activities. A cautionary suggestion was put forth by visiting American industrialists that in such situations the requirements for workforce skills and for efficient facilities may be so different that sometimes the best approach is simply to walk away from a defense facility and start over with a different type of plant and new workers. The Soviet official was clearly not prepared to accept this suggestion which would further complicate his task of formulating general policies for conversion programs.

One of the most important management skills in the USSR, at both the national and enterprise levels, is the ability to ensure a steady flow of critical materials and supplies required for production in a country where there are not enough materials and supplies to go around. The objective is not to achieve the success of Japan in having materials and components arrive only a few hours before they are needed and thereby reduce inventory costs; the problem is much more basic, namely, to ensure that materials vital to keeping the production lines flowing are on hand at any cost, even if they have to be warehoused for many months.

We have already looked at how the planning process in Moscow accords first priority to the production needs of the military sector and how national delivery schedules for materials and supplies for military programs are developed accordingly. Sometimes, these delivery schedules are not met, despite the enormous planning effort and their priority. When shortages occur, however, the managers of military facilities can call on local party officials where the production facilities are located to intervene through their channels. These officials do not hesitate to arrange for the prompt diversion of connectors, cabling, and other items from the television factory, for example, to the missile plant. Similarly, military production managers, reinforced by party officials, have little difficulty in emphasizing to their suppliers the importance of high-quality materials and components that meet military specifications. In short, when the military establishment speaks, suppliers listen.

The new management teams transferred to the civilian sector from the defense sector are now experiencing a rude awakening. Ensuring on-time delivery of high-quality materials for civilian factories is not so easy. Regardless of the personal connections and clout of the managers, materials that don't exist in the civilian sector cannot be delivered. Further, the many people involved in the procurement process in the civilian sector have become so accustomed to a sluggish system that doesn't work very well that inspiring them to more determined action is a Herculean task.

An important aspect of Gorbachev's perestroika is the outreach to the West for technologies which can truncate the economic development process. Joint ventures, licensing agreements, and cooperative programs with foreign firms are being pursued as never before. Similarly, in the area of basic science, many Soviet organizations are pressing for East-West exchanges, international collaboration on global problems, and cooperation on big science projects which exceed the financial resources of any single country. The Soviet leaders grasp the potential significance of the rapid spread of science and technology throughout the world for advancement of the Soviet society. They clearly intend for the USSR to become an active participant in this diffusion as the only alternative to continued technological backwardness.[3]

*   *   *

The reduction of the size of the Soviet bureaucracies is underway—painfully, sporadically, but persistently. Already many senior officials have retired, and more are scheduled to become pensioners every month. Many younger office workers have been dispatched to middle-level management jobs in industrial enterprises. Frequent Western visitors to government offices in Moscow regularly encounter new faces at the table with no information volunteered concerning the fates of their predecessors.

Some of the former bureaucrats who now spend their time in their gardens rather than their offices are quite bitter about this treatment, and particularly about their loss of power and influence. For a brief time they try to use their old-boy networks in rebelling against the system, but usually to no avail. However, other early retirees have accepted their loss of bureaucratic power as simply fate and are rapidly fading into the countryside.

The younger generation of displaced bureaucrats is accustomed to being programmed, and they believe that the system will take care of everyone. Indeed, this usually has been the case. Somehow large numbers of displaced, young white-collar workers find new jobs in the same geographical locations. Earlier suggestions that new jobs awaited them only in distant cities of Siberia apparently have remained simply suggestions.

Before the advent of perestroika, 15 million government administrators served throughout the country. About 3 million of these were layered at the upper level of the bureaucracy in Moscow and in the capital cities of the republics. Gorbachev aims to eliminate one-third of the 3 million jobs. The staff of the State Committee for Science and Technology of more than 2000, for example, is being reduced by more than 30 percent. Cries of anguish emanate from the department chiefs of the committee who lose not only their staffs but also much of their office space. Comparable targets for staff reductions have also been set for the many other government agencies responsible for the administration of scientific programs.

*   *   *

With regard to the evolving private sector, a few ingenious Soviet entrepreneurs will undoubtedly become wealthy in the months and years ahead, as was the case during the Soviet experiments with private enterprise in the 1920s. The Soviet leadership is clearly concerned about the societal implications of the private cooperatives springing up in all cities which permit private entrepreneurs to pool resources and share profits. By the end of 1988, over 700,000 people were working for cooperatives in the industrial and service sectors; leading Soviet economists have projected a tenfold growth of cooperatives during the next decade, but at that time they still will account for only about 5 percent of the workforce. Plans are being formulated, for example, for a new cooperative in Moscow to manage the country's largest hotel and to build and operate a steel mill; another Moscow cooperative now directly and indirectly employs 2500 people. In the Soviet republic of Georgia, the feasibility of a new internal airline managed by a cooperative is being considered. The government initially placed a limit on the number of employees in each cooperative and has been trying to establish a tax structure which will provide some control. However, the specter of large profits excites many talented Soviets to devise ways to circumvent such controls. At the same time, most Soviet citizens even under perestroika have great difficulty coping psychologically with the notion that a handful of Soviet citizens may become very rich.

One widespread hope in the USSR is that cooperatives will serve as agents for transferring technology which is currently isolated in a few state enterprises into other parts of the economy. To date many cooperatives have been established which rely on moonlighting engineers to provide the materials, equipment, and services for the repair of heating, plumbing, and electrical systems in both industrial and residential facilities. Sometimes they share profits with state enterprises in exchange for access to needed materials and equipment. Initially, these types of cooperatives worked primarily for private customers, but they increasingly receive contracts from state organizations as well. This low level of technology diffusion is desperately needed in the USSR.

One extraordinarily successful cooperative which is composed of moonlighting scientists from an electronics research institute has succeeded in producing with a small labor force high-precision lenses using materials that cost 100 rubles per lens. The factory which has been producing this type of lens is either so inefficient or so clever that it has been selling its product to a state trading company for 10,000 rubles per lens. The trading company is now buying the lenses at the higher price from both sources. The scientists are delighted, the factory is embarrassed, and the government does not know what to do.

Cooperatives devoted to computer software pique both military and consumer interests. Will engineers from the defense sector be able to moonlight? Will such cooperatives be able to link with Western partners? Who monitors the software products to exploit them for military applications?[4]

*   *   *

Soviet intellectuals, and particularly scientists and economists from the Academy of Sciences of the USSR, are playing a pivotal role in developing practical solutions to the systemic and technical problems which block Soviet industrial progress. Some of these recent allies of the party leadership were lonely critics of the shortcomings of the approaches of Gorbachev's predecessors, and they are now trying to energize the entire society through support of perestroika. They are studying Western management techniques. They are preparing the nation to use computers. In many ways they are manning the tiller of a ship in its earliest stages of renovation as it sets out prematurely but necessarily into unfamiliar waters. Never before has such a heavy burden been placed at the doorstep of scientists in the USSR or anywhere else.

It is not surprising that Gorbachev has repeatedly called upon Soviet scientists and other intellectuals for support and personal counsel and that many leading Soviet scientists have embraced his approach. In contrast to his predecessors who had little formal education, he feels completely at home with the intellectual community, having graduated from the law faculty of Moscow State University. He seems to thrive on the controversial debates that are now adding spice to party meetings, and he relishes encasing his pragmatic program pronouncements within sophisticated conceptual assessments.

Gorbachev has great hope in the potential of science. As already noted, he witnessed the dismal failures of poorly conceived agricultural policies while science was contributing to spectacular agricultural successes abroad. At the same time, he has been mightily impressed by the Soviet achievements in the fields of military and space technologies.

The Academy of Sciences has become a focus for many of the hopes and the frustrations of the Soviet leadership. At its formal sessions, the Academy engages political leaders in debates of the most pressing issues of the day. Gorbachev in turn invites selected academicians to partake in debates within the upper echelons of the party. The Academy has established advisory commissions on issues ranging from prevention of nuclear war to the cleanup of Chernobyl to protection of the ecology of Lake Baykal to reconstruction of Armenian cities. Moreover, many Soviet academicians are regularly consulted by government officials on energy, health, and other major issues.

Similar to practices in the West, the advice coming from the Academy and from its individual members is not always the type of advice being sought, however. Academicians have been particularly critical of what they consider to be outmoded approaches to education and of the general lack of ecological concerns by Soviet industry, for example. While the institution remains cautious in its criticisms of the Soviet leaders of the past and present, it has joined the chorus of public criticism of fundamental flaws, particularly the emphasis on quantity rather than quality, in an economy that only responds to commands from a central authority.

*   *   *

The geriatric character of the Academy of Sciences of the USSR has been a target for perestroika. Surprisingly, in 1987 the governing body of the Academy, the presidium, accepted without great controversy the concept of mandatory retirement of Soviet scientists from senior management positions at age 65. As exceptions to this rule, the 400 scientists who have been elected to full membership in the Academy can continue as managers until age 70, and the 43 members of the Academy's presidium until age 75. When the vote was taken, many presidium members were in their 70s and 80s.

Prior to being asked to vote, the presidium members were assured that there would be no loss of financial benefits for retirees. They would retain their full salaries, their chauffeurs, their country homes, and their access to laboratories and support staffs. The general consensus of the presidium members was that such a radical proposal with its substantial financial implications for the government would never be implemented regardless of their vote. Therefore, they saw no need to oppose the latest "policy fad." How wrong they were.

Now the retirement policies are in effect within the Academy system. Institute directors are being replaced. Younger scientists who are in touch with the political and technical realities of the day have been elected to the presidium and are having a major impact on developments within the Academy. The Tuesday morning meetings of the presidium have even taken on a lively atmosphere as participants vigorously press their often conflicting viewpoints.

Related to mandatory retirement is a recent change in the former policy that new members could be elected to the Academy only when positions opened up due to the death of members. Now academicians are considered to have passed on—for the purpose of opening positions for new members—when they reach the age of 75, although they retain their status as academicians. When this rule became effective in late 1987 more than 100 academicians were elected to the Academy, increasing the number of full members by 30 percent and reducing the average age of Academy members by more than seven years.

Another proposal by several leading academicians to bring fresh blood into the management of Soviet research programs calls for directors of Academy research institutes to step down after two five-year terms. The proposal has not been accepted by the Academy, however. As noted in the previous chapter, many institute directors have settled into these comfortable and powerful positions for the lifetimes of their working careers. Directors have commonly remained in place for 20 or more years. In many cases institutes with staffs numbering hundreds of scientists become highly personalized operations, with only the director and his closest friends even aware of many of the activities of the institute. However, reform may gradually come into effect as five-year terms expire and some incumbents agree to step aside in the spirit of perestroika.

Pay scales have long been criticized by Soviet researchers and particularly outstanding young researchers. Beginning in 1985, salaries for scientists have been on the rise—within the Academy, within the educational institutions, and now within the institutions of the various ministries. An important step has been to establish additional pay levels for researchers that enable them to progress higher up the salary scale without moving out of research and into management. In very general terms, starting pay levels for young scientists are comparable to pay levels for skilled workers and other professionals—up to 200 rubles per month. However, promising scientists have a chance for early increases, and within a few years they may be earning 400 rubles, which is generally more than the salaries of counterparts in many other professions. While relative salaries among workers, scientists, and other professions will always be a contentious issue, an improved framework for compensating researchers is now developed.

For many years the government has urged Soviet researchers to come up with new inventions and with patents that will make a difference for Soviet industry. An elaborate system of financial and political rewards is in place. However, the committees to recommend rewards frequently have difficulty finding enough applications to make the system function. Indeed, as previously underscored, encouragement of individual creativity has never been a strong dimension of the Soviet research system. Most scientists are housed in large institutes which are structured in a hierarchical way, with successive levels of management having increasing control over the research activities. For those achievements deemed very important, the institute directors or the deputy directors quickly become the focal points of the efforts. For achievements of secondary importance, the division and laboratory chiefs take charge, at least in name. The junior researcher may have difficulty even publishing his or her most important findings, let alone sharing credit for any breakthroughs.

The Academy of Medical Sciences of the USSR recently borrowed a concept from the West and established a competitive grants program. Individual researchers prepare proposals and compete for funding for their own projects. In the first round of competition, researchers from a variety of institutions submitted about 400 applications—impressive for the Soviet Union but very modest for the United States, where many scientists build careers around grant proposals. About 100 Soviet scientists received grants, totaling approximately $1 million in hard currency, which can be used to buy equipment and supplies from the West to support their research activities. Given the difficulty for Soviet scientists to obtain foreign currency and the underdeveloped state of the Soviet industrial capacity to produce reliable scientific instruments, these funds are very important to the successful applicants.

A competitive grants program has long been under discussion within the Academy of Sciences of the USSR but has not yet been adopted. A few senior Soviet academicians familiar with such programs in the United States repeatedly press for such an approach. The president of the Academy has expressed concern that the proposed diversion of funding from support of projects developed by institutes to projects developed by individual scientists could cause many problems for the institutes and weaken their overall programs. According to other Soviet scientists, this reluctance to aggressively pursue a policy of competitive grants for individual scientists is attributable to "conservative" forces within the Academy which prefer research projects to be planned from the top rather than initiated from the bottom.

While programs of individual grants might be a step toward encouraging individual creativity, many institutions in the USSR simply are not prepared to release their most valuable researchers for work on projects that are not the top priorities of the institutions and to provide the required logistical support for such individual activities. I had related experiences as the director of a US Government laboratory. Our laboratory received substantial government funding each year largely because of its reputation of being able to provide useful research results of interest to government agencies within a reasonable period of time. At the same time, many of our best scientists wanted to have their personal research grants so they could work on their own and search for the "real breakthroughs." In a university setting, such searches for the unanticipated are to be applauded; but in a government laboratory such diversions can undermine the entire funding base of the laboratory. Institutes of the Soviet Academy have historically had characteristics of both US Government laboratories and our university laboratories: they are responsible for exploring the unknown, and in this regard their scientists should have greater freedom in choosing their own projects; but they also have shouldered a heavy burden to respond to priorities established at the top. Given the inevitable conflicts between these two responsibilities, questions are frequently raised within the USSR of the future viability of the Academy system as it is currently structured.

Increasingly, young science "superstars" are beginning to have the opportunity to emerge from the lower echelons of the Soviet institutes, usually under the personal patronage of Academy members. An important aspect of recognition of young talent is the increasing openness of Soviet institutions to foreign visitors. Many visiting scientists who meet promising young researchers in Soviet laboratories do not hesitate to bring this potential to the attention of their senior Soviet colleagues. Foreign recognition of Soviet scientific achievements is very flattering, and Soviet administrators pay attention to unsolicited comments by visitors. These administrators are now responding more positively to suggestions of greater international exposure for young Soviet scientists through participation in international conferences.[5]

*   *   *

As reviewed earlier, education is big business in the USSR, with over 55 million people enrolled in some area of formal education. A high degree of standardization and rigidity compensates for the absence of an adequate supply of well-trained teachers. The lack of a well-developed printing industry in the USSR greatly hampers preparation of a diversity of up-to-date written materials for classes, adding further pressures for standardized approaches. At the same time, reformers call for education tailored to the needs of individual students, needs which depend both on the capabilities of the students and on employment requirements after graduation. However, significant movement toward individualized education on a large scale seems unlikely. Centrally programmed approaches must persist for the foreseeable future, with the primary focus of reform targeted on the standard curriculum.

Education officials recognize the necessity for upgrading the cadres of teachers, and particularly science teachers, at the primary and secondary level. Salaries have been raised, although they remain low in comparison with other professions; a factory worker often receives 20 to 40 percent more than a secondary school teacher. Retraining of teachers who are out of date with recent trends in the education field is also expanding.

No educational program has received more publicity in the USSR in recent times than the introduction of computer instruction into the secondary education system during the past several years. The concept of computer literacy training for every student is certainly laudable. However, comments in 1988 by a realistic Soviet scientist were quite telling: "Learning computer skills from a book without access to a computer is like learning to ride a bicycle without access to a bicycle."

The Soviet Government has ambitious plans for equipping schools throughout the country with personal computers, and now a few schools in every major urban area have computers for classroom instruction. Also, the government has formed computer clubs in Moscow and other large cities where young Soviets can spend their time after school hours. The size of the effort to date is small, and the quality of the computer facilities is modest, but the enthusiasm of the few participants is very high. As in all countries, skills in the computer field are considered a passport to desirable employment. One advantage that Soviet youths have in the field of computer science is the strong emphasis on mathematics at all levels of the Soviet educational system. Thus, they develop software skills with relative ease.

Illustrative of the responses of Soviet scientists to their frustrations over the slow progress in advancing computer literacy is the program of computer education for secondary school students in a suburb of the Siberian city of Novosibirsk. This suburb, named Akademgorodok, or Science City, is home for many scientific research institutes. A Soviet academician from the Academy's computing center has succeeded in equipping a number of local schools with personal computers. While visiting some of these schools in 1988, I was very impressed with how he had purchased so much equipment by obtaining funds from the budgets of research institutions in Akademgorodok. He had then taken the lead in introducing computer courses into the curriculum for several years before his recent death. While he was greatly admired throughout the USSR for his initiative and determination in putting this program in place, some education officials in Moscow question the pedagogical basis of introducing computers on a piecemeal basis prior to development of an overall educational approach for computers which is integrated into the more general curriculum. Perhaps at the outset, any type of computer instruction is better than none; and in the years ahead the education officials will have a chance to build on his pioneering efforts.

Turning to research at the Soviet universities, or perhaps more correctly the paucity of research, the highest party leaders have repeatedly called for greatly expanded cooperation between the science departments of the universities and the research institutions of the Academy of Sciences and of the ministries. In practice, more of the bright science students who favorably impress their professors should have an opportunity to carry out course work at research facilities of the Academy and ministries. Also, the number of graduate students working for their advanced degrees at Academy institutes should increase.

Universities are expected to raise research funds through contracts with industry in an effort to link research and education more closely together. A few talented professors and even graduate students will undoubtedly have new opportunities to supplement salaries and stipends through such contracts and consulting arrangements. However, universities in the USSR simply do not have much to offer industry in the way of research that has short-term payoff.

As already noted, the education and research programs of the many engineering colleges in the USSR are usually of a highly applied nature. While Americans tend to consider most technical colleges in the United States as a cut below our universities, this is not the case in the USSR, where a number of engineering schools stand out. It has been said that Soviet society was built on the backs of its engineers, who led the early industrial drives of the 1920s and 1930s. Most of these engineers were products of the engineering colleges.

The linkages between the engineering colleges and industry are often strong, and industry continues to draw its new engineers from such institutions. These institutions are often equipped with engineering hardware similar to equipment used in industry, and the college faculties and students have the luxury of time not available to those in industry to experiment with new ways of using the equipment. As a new step to strengthen the link between industry and the engineering colleges, industry is being asked to pay fees to the colleges to hire graduates, with the fees helping to subsidize research. Industry, in turn, is taking a greater interest in the direction and quality of the research that it subsidizes.

Many years, and probably decades, will be required to bring about fundamental changes in the nature of the Soviet educational system. New facilities, new generations of teachers, and new materials and equipment are needed on a massive scale. On the other hand, the organizational and personal links between educational programs and activities in other parts of the scientific community should improve substantially as the government encourages freer flow of information throughout society.

As in all countries, the challenge is to preserve the best of the educational system while strengthening or replacing the weak links. The basic problems in the USSR are rooted in decades of financial deprivation of the educational sector, in an educational approach that reflects the central planning philosophy of the Communist system, and in historical compartmentalization of intellectual processes by discipline. Many of the excellent science and engineering students will continue to thrive regardless of the system. Somehow, they will find the right textbooks, they will link up with leading experts, and they will find time on the computers. The real challenge is to raise the level of opportunities and expectations for a broader range of students.

*   *   *

The ever-present concern over translating research into practice is at the heart of the science reforms. In every country technical, financial, and other hurdles must be overcome in moving an idea from the laboratory to widespread industrial practice. In the compartmentalized Soviet economy the problems are extreme. The reform program in theory and in practice concentrates on improving the delivery systems for bringing the results of successful research with potential near-term applications into the production and service sectors of the economy as quickly as possible.

How is this to be done? The enterprise managers contend that the scientists have failed in the past—their discoveries are few and far between and with an occasional exception are irrelevant to the needs of the industrial organizations. The scientists argue that the significance of their efforts is not appreciated, that they do not receive adequate support to take their research to completion in a timely manner, and that the enterprises are not sufficiently enlightened to change their ways even when presented with new technological opportunities. The bureaucrats often throw up their hands and say, "Bring in the Germans or the Japanese."

New organizational structures have been devised to place research and production under the same management team, an approach commonly encountered in American industry. New long-term planning mechanisms are to ensure that investment funds will be available when technologies emerge for improving industrial processes on a nationwide basis. Also, financial rewards are promised for scientists and plant managers who successfully introduce technologies that improve efficiencies.

Other new management, financial, and organizational ideas to link research and applications are suggested every week—at party meetings, at scientific conferences, in technical journals, and in the press; and many are in place. Academicians, engineers, and economists now head large complexes that include research facilities, design bureaus, experimental factories, and production facilities. Many temporary research laboratories have been established to solve very specific problems facing industry. Researchers have part-time assignments to work in the enterprises, giving advice on how to upgrade manufacturing processes. Research institute budgets must be shored up with contracts with industry, thereby forcing close institutional contacts as a basis for survival. Administrators are increasingly siting research institutes immediately adjacent to industrial complexes.

At the individual industrial enterprises, the plants are expanding technical staffs, and exploration of research opportunities within the factory boundaries is gradually becoming a legitimate industrial activity. Political exhibitions of industrial achievements regularly require displays by the enterprises showing how they have introduced innovations and how modern technology has raised productivity. Special funds of the enterprises support research investments.

Everyone seems to be keeping score as to how many innovations are introduced into practice each year. The research institutes list all of their developments that can be seen on the factory floors, and each ministry and the Academy of Sciences keep master lists of how they have put science into action.

While these approaches will help improve the linkages between research and production, some fundamental problems will persist. Most production personnel and research scientists have simply operated on two different planets. Traveling from one to the other is indeed a galactic voyage. Job mobility between these two spheres is virtually unknown. Exchange of technical information is often equated with suspicious behavior. Also, each employee is expected to complete his or her own task, and the results then become someone else's problem. Any proprietary interest of researchers in trying to guide the result of their research as it makes its way toward application can only lead to trouble for the researchers, so they believe.

Until such travel from the research setting surrounded by scientific publications to the domain of industry, where fulfillment of production quotas still rules the day, becomes routine for many technical specialists, the separation of research and applications will continue. In the United States we have product champions who superintend research ideas from their earliest experimental phases until they are earning money in the marketplace. Such follow-through is sorely lacking in the USSR.

Many years ago the Soviets established a system of centralized research institutes to service each branch of industry. A few of these institutes have always had close working relationships with some of the enterprises that they serve, and particularly enterprises in the Moscow and Leningrad areas. Naturally, these institutes strongly object to the insinuations that there exists a decoupling between research activities and the interests of the enterprises. Yet, two fundamental weaknesses stand out. First, the institutes are so attuned to the near-term problems of the enterprises that their activities are usually more of a service than research nature. While such services are important, they are not a substitute for research into new and better ways to produce industrial products. Second, some enterprises are so far away from the institutes—often more than 1000 miles—that they seldom feel the impact of the programs of the institutes.

In order to create effective coupling between research institutes and industry, institutes must have something to offer. Their technical staffs must understand the industrial processes as well as or better than the staffs of the enterprises. Their research results must indeed be relevant and hold high promise for improving products or processes. The process of adaptation on the factory floor must be within the realm of the possible. In the past many Soviet research institutes have had little to offer other than talk.

While the Soviet craving for usable research products is intense, Gorbachev and other political leaders have also accepted the idea that fundamental research to explore the unknown irrespective of potential near-term applications should not be ignored. The president of the Academy of Sciences has noted that the Soviet effort in fundamental research that does not have immediate application is only one-fifth that of the American effort. According to his statistics, twice as many American scientists are engaged in fundamental research, and they are almost three times more efficient given the modern equipment at their disposal.[6]

Meanwhile, the directors of many Academy institutes and some institutes of the ministries are rebelling against what they consider to be an excessive emphasis on contracts with industry which distort their research plans. Within the Academy the institutes seem to be enforcing an upper limit of 30 percent of their budget coming from industrial contracts, and the Academy seems to be willing to continue to subsidize the remaining research ranging from 70 to 100 percent depending on the type of research of the institute. Clearly industry will be more interested in contracts with an institute for electrical engineering than an institute for botany.

I recently visited a Soviet drug research institute that was required for the first time to identify the drugs it would develop before it received its annual budget allocation. Previously, the institute simply had a general fund to pursue research on drugs in any fashion it desired. The institute was unaccustomed to such advanced planning and accountability. Desperate to respond quickly to its instructions from the Ministry of Health to detail its research plans, the institute simply selected 100 drugs which were sold in large quantities by Western companies but which were not available in the USSR. The institute then asked the Ministry of Health to select from these drugs the most interesting ones, which the institute would replicate for the Soviet market. Given the underdeveloped state of Soviet pharmaceuticals, this approach is understandable. However, should Soviet research talent be directed simply to copy Western products? Directives from the ministry should be designed to encourage the creativity of individual scientists, not to force the scientists simply to rely on the work of others.

*   *   *

While the outward appearance of the buildings, the contents of the shop windows, and the availability of consumer products within Soviet cities have changed very little during the past decade, the recent changes in attitudes, policies, and plans have been profound. Glasnost and perestroika are indeed making their presence felt throughout the country, and particularly among intellectuals within the scientific community.

Freedom of scientific and political expression and the delegation of control of research activities to the levels of the institutes and the laboratories are becoming realities. More and more scientists have stronger voices in their future as many of the older, autocratic generation leave the scene. But scientists are realists and do not expect the entire approach to research to change overnight. However, they will surely resist a return to the straightjackets of the command and control approaches of the past.

The problems facing Soviet science are formidable. Until the formal and informal information systems that characterize science the world over gain significant strength in the USSR, Soviet scientists will have great difficulties staying abreast of developments at home or abroad. Large investments in facilities and equipment are required to let Soviet science reach the world levels so often mentioned. Most importantly, scientific creativity cannot be stimulated overnight, and Soviet scientific leaders must be patient.

Revamping the educational system will also not be easy. Hundreds of thousands of teachers and professors at all levels are not eager to see the programs that they have developed replaced by new approaches yet to be tested in the USSR. The educational establishment has adopted a siege mentality, and many of the leaders feel they have not received a fair hearing. The old guard is unimpressed by the call for new approaches and for innovations such as computer education without computers. Still, the academicians who have joined the struggle for educational reforms enjoy tremendous prestige among the public. With easy access to the media, they press their cases for reform.

Turning to industry, the Soviets will be hard pressed to enter the twenty-first century with the high-technology strengths of West European or many of the East Asian countries. Organizational fixes and turnkey plants from the West will undoubtedly be emphasized for the next few years. Some inefficient Soviet enterprises will probably fail as the concept of bankruptcy gains legitimacy, and a private sector may begin to take hold in a few areas of interest to Soviet consumers—clothing, building maintenance, vehicle and equipment repair, publishing, health services—but not in those of critical importance to the functioning of the Soviet economy.

The Soviet model for future industrial growth has yet to be defined. It will not be patterned after the renowned Hungarian model or any other Socialist model nor will it resemble the Western capitalist models. The USSR is simply different—in its natural endowments, in its ideological commitments, in the talents and interests of its people, and in its leadership. The Soviet model will be unique. It will undoubtedly be designed to ensure the continuation of the USSR as a world power under a changing definition of power that recognizes the importance of economic prosperity at home as well as military strength. Soviet scientists will be expected to play central roles in promoting this economic prosperity while maintaining the military capabilities.

CHAPTER 5

Scientists March to the Diplomats' Drums

Man is by nature a political animal.
Aristotle

We are handicapped by [foreign] policies based on old myths rather than current realities.
Former US Senator J. William Fulbright

The adversarial relationship between the United States and the USSR since World War II has led to careful control by the two governments of interactions between American and Soviet scientists. This control has affected all aspects of cooperation, from the initial negotiations of scientific exchange agreements to governmental debriefings of scientists after visits abroad.

In recent years detailed negotiations have usually been the first step in establishing serious cooperative programs involving scientists and engineers from the two countries. In most cases, governmental representatives from the two countries have conducted the negotiations. In a few instances, and particularly during the past several years, private American institutions have carried out their own negotiations with Soviet organizations—organizations which of course are governmental.

Negotiations may last a few days, may extend over many months, or may even continue during a period of years. The negotiators usually address every word and every comma in the proposed agreements, memoranda of understanding, and implementing protocols. Formal documents cover planned activities in excruciating detail. They usually establish quotas for person-months of exchange visits, set forth the scientific fields which are appropriate for cooperative activities, and specify the financial details including the per diem allowances. They establish deadlines for exchanging correspondence and timetables for implementing exchanges. They may address protection of patents and copyrights and sometimes specify the reports which are to be prepared.

While a diplomat assigned to the American Embassy in Moscow in the mid-1960s, I experienced the painstaking aspects of negotiating formal agreements between the two governments. Initially, I witnessed the final stages of the negotiations of a consular agreement which led to the opening of a Soviet Consulate General in San Francisco and an American Consulate General in Leningrad, negotiations which had lasted 30 years. Then I was a participant in negotiations of an intergovernmental exchange agreement between the two countries in the fields of education, culture, and science.

Our negotiations of that exchange agreement took place during a cold Moscow winter. They lasted eight consecutive weeks as we huddled each weekday in heavy sweaters and coats in a poorly heated conference room at the Ministry of Culture. Most of the text of the agreement had been worked out in advance of the negotiations. The remainder required about two weeks of discussions concerning several difficult points and some of the details of carrying out exchanges; the appropriate resolution of these issues rapidly became clear. But the political setting was not right to reach final agreement. So we continued to meet, but we waited; and we waited. The end came quite abruptly at a luncheon on the fifty-third day after both delegations received almost simultaneously instructions from their governments that the time was appropriate for a significant political step to help reverse the deterioration of bilateral relations. In short, the major disagreements were resolved without delay, and compromises on details were easy to devise. However, the political will was the critical ingredient.

As another example, in the summer of 1985, I had the task of negotiating in Moscow the details of a scientific exchange agreement between the National Academy of Sciences and the Academy of Sciences of the USSR. The National Academy is a nongovernmental nonprofit organization chartered by Congress to provide advice to the US Government. It also conducts a large program of international activities which includes scientific exchanges with many countries. Its program of scientific exchanges with the USSR dates from 1959. In January 1985, the leaders of the two Academies had agreed on a lengthy set of general principles to provide the framework for the agreement. While there was continuing Soviet reluctance to stick with these principles, in the end we incorporated them fully into the agreement.

Our negotiations lasted only three days, with each day punctuated by an alcohol-free luncheon in accordance with the new policy of Gorbachev. The Soviets had five negotiators who had many difficulties agreeing among themselves. Both sides were determined to conduct the discussions expeditiously and without polemics; and during our word-by-word review of the texts prepared by the two Academies, we quickly resolved all but three issues. We agreed to refer to the presidents of the two Academies the precise formulation of clauses concerned with human rights and with the role of individual scientists in selecting their collaborating partners in the other country.

Also, the Soviet financial authorities needed time to determine how much support they could provide for dependents accompanying American exchange scientists to the USSR. Within a few weeks these issues were resolved by correspondence, with both sides ratifying language which had been worked out in Moscow.

Since the National Academy of Sciences is not a US governmental institution, I was far less constrained in 1985 than in 1965. Also, the leaders of both Academies had set the policy tone in their agreed principles, and we were therefore concerned primarily with elaboration of these policies and the mechanics of implementing exchanges. Nevertheless, telephone calls frequently interrupted our meetings as the Soviet Government kept in close touch with the negotiations, particularly as we crafted language on human rights issues. From our side, I was not operating under governmental instructions in any sense. At the same time, our Academy wanted to develop a document that was compatible with provisions being proposed by the US Government for inclusion in an intergovernmental cultural exchange agreement under negotiation at the same time. The National Academy also needed to be sensitive to concerns within the American scientific community over human rights issues and the Soviet preoccupation with secrecy in limiting the flow of information between Soviet and American scientists. More specifically, one negotiating objective was to ensure that the agreement facilitated rather than impeded access by American scientists to Soviet colleagues in the USSR; and we stood firm in rejecting standard Soviet language that could complicate such contacts, namely, "will comply with the regulations of each country" and "will not interfere in the internal affairs of the host country." Internal security controls were tight in the USSR, and such phrases could provide additional reasons for Soviet crackdowns on contacts with foreign scientists.

*   *   *

The US Department of State and the USSR Ministry of Foreign Affairs maintain general oversight over formal and informal scientific and technical contacts. In the USSR, the State Committee for Science and Technology also plays a key role in arranging and watching scientific exchange activities, and it has had strong links to the intelligence services in the country. Domestic and international political considerations influence the policies of all of these organizations toward exchanges. This political aspect of US-Soviet exchanges is in stark contrast to the tradition of minimal influence of the US and other Western governments on the travel of scientists and the open sharing of scientific information and facilities among most countries.

Many other governmental ministries, departments, and agencies within the two countries have major interests in scientific exchanges. At the highest levels, decisions concerning exchanges are sometimes made by the White House and the central committee of the Communist party after internal debates among politicians, diplomats, generals, and scientists. These high-level decisions have recently included the approval of new intergovernmental agreements, for cooperation in basic research and in space sciences for example, and review of highly visible cooperative projects such as US and Soviet involvement in multilateral cooperation in fusion research. In the United States the White House sometimes reviews requests for governmental funding by private organizations interested in specific exchanges, as illustrated below.

Proposals for exchanges often originate within the technical agencies of the two governments—the agencies responsible for energy development, space exploration, environmental protection, geological studies, and basic science. These agencies negotiate the details of the cooperative programs, and they carry out or fund most of the exchanges. In the United States, during times of political stress between the two countries, the agencies try to maintain momentum in existing programs but are hesitant to make new proposals for cooperation lest they appear to be meddling in the foreign policy arena. During the times of détente, they may seize the moment to press for favorite additional activities. The agencies clearly are responsive to the political process.

Representatives of the intelligence agencies in both countries argue within their governments over proposed cooperative programs. They seek opportunities to obtain information about an adversary, and from that vantage point they favor exchanges. At the same time, they are concerned that exchanges could provide the same adversary with comparable opportunities. Sometimes they brief exchange scientists, particularly scientists from government agencies, before their visits and debrief them after the visits.

However, the KGB has been much more aggressive in influencing exchange activities than have US intelligence and counterintelligence agencies; as we will see later, this aggressiveness is attributable in large part to the role of the KGB in collecting technical information for both military and civilian applications. American agencies do not have similar responsibilities since we rely to a large extent on the normal commercial contacts of American companies to stay abreast of worldwide technological developments. Indeed, the roles of the intelligence and security agencies in the two countries are simply not comparable.

In the United States many political personalities, particularly congressmen, press for their personal agendas for bilateral programs. One currently popular theme is cooperation in the environmental field; and many Congressmen repeatedly urge the Executive Branch to expand activities in this field. Notwithstanding such pressure, the Department of State continues to play a central role in the determination of the extent and character of cooperation and the appropriateness and timing of specific exchanges. The Department of State usually defers to the technical agencies once exchange programs are in place, but occasionally foreign policy considerations override the preferences of the agencies. The State Department has the most comprehensive overviews of interactions between the two countries at any given time as well as the legal authority concerning international activities, and, therefore, is able to mount persuasive arguments favoring or opposing exchange proposals.

Very importantly, the Department of State and the Soviet Ministry of Foreign Affairs control the issuance of entry visas for visiting scientists, with each issuance being decided on a case-by-case basis. These decisions are often made just prior to the visits; while such last minute visa decisions are frequently due to the administrative delays in the system, they are repeatedly interpreted by the affected scientists as being motivated by foreign policy or internal security concerns. There is no more powerful reminder of the governmental control being exercised over exchanges than the uncertainty in the mind of the American or Soviet scientist as to whether a visa will be issued on the very eve of departure to participate in an exchange.

From time to time, American scientists suggest new types of East-West cooperation that raise political concerns, such as cooperating with Soviet scientists in politically sensitive Third World countries. Also, with glasnost Soviet scientists may propose cooperative programs that have not been scrutinized by their government. Still, the launching of major cooperative ventures by public or private institutions without the active support, or at least the acquiescence, of both governments is highly unlikely.

*   *   *

In view of the key roles played by the foreign affairs organizations of the two countries in cooperative endeavors, exchanges are inevitably entwined in foreign policy deliberations. In some cases, scientific cooperation has provided the concrete evidence of changes in foreign policy. And in many ways, the level of intergovernmental exchanges in general, and scientific exchanges in particular, has become a bellwether of the state of political relationships between the superpowers.

As summit meetings of the leaders of the two countries approach, officials in both countries traditionally search for areas of scientific cooperation which can be incorporated into new agreements, communiqués, and joint statements. Heads of state enjoy announcing areas of agreement which will benefit mankind, even if the announcements simply expand programs which are underway. As previously noted, at the 1985 summit in Geneva, calls for expanded cooperation in fusion (which began 30 years ago) were to signal an eventual end to energy shortages. At the 1987 summit in Washington, joint research in the area of climate change (which had been underway for several years) was to help ensure that the hole in the ozone belt over the Antarctic region would not spread. In 1988 in Moscow, the leaders of the two countries trumpeted cooperation in the exploration of Mars (which had been discussed for almost 20 years) to open exciting new horizons.

In the early 1970s, Secretary of State Henry Kissinger launched a decade of expanded bilateral scientific and technological cooperation as one of the centerpieces of US efforts to improve relations between the two countries. This cooperation was brought to life in 11 formal intergovernmental agreements in science and technology. The direct costs to US agencies for these overtures reached millions of dollars annually, as many hundreds of scientists from each side made trips abroad to plan and implement cooperative efforts. These programs had the desired political effect of translating the concept of détente into highly visible terms. They improved mutual understanding of the relative technical capabilities of the two countries. In some cases they resulted in discernible advances in science and technology. At the same time the cost of this cooperation was insignificant in comparison with the expenditures on science and technology directed to military confrontation—millions versus tens of billions of dollars.

Conversely, in times of political controversy, scheduled scientific exchanges become an easy target for postponements to demonstrate unhappiness with the policies of the other country. For example, during the early 1980s, four agreements in space sciences, energy, transportation, and other areas of science and technology were allowed to lapse by the United States in response to Soviet actions in Afghanistan, in Poland, and elsewhere. Also, Soviet destruction of the Korean airliner 007 and harsh Soviet treatment of political dissidents led American officials to slow down cooperation under other agreements.

*   *   *

In 1984, President Reagan began his search for new exchange programs to improve US-USSR relations, which eventually led to the announcements of scientific cooperation at the summit meetings. While arms control clearly was the major focus of Reagan's efforts to improve relations, scientific cooperation served as a useful political complement to arms control initiatives with uncertain outcomes.

Spokesmen for the Reagan administration disavowed what they called the "Kissinger approach" of promoting scientific and technical exchanges primarily for foreign policy reasons. They began to emphasize that scientific merit was to be the overriding criteria in deciding whether to move forward with future cooperative arrangements. However, the internal debates in the mid-1980s in Washington over scientific cooperation with the USSR were often politicized. Arguments over possible future military applications of science were always on center stage. Conservative forces unwilling to admit that the United States might obtain useful information or that science might benefit from cooperative programs frequently dominated the discussions. They were determined to avoid the possibility that the Soviets might benefit in any way from scientific and technological exchanges.

In 1986, the White House overturned an earlier decision of the National Science Foundation, a US Government agency, to accept a contribution of $2.5 million from the Soviet Academy of Sciences and to include Soviet specialists in a geological program for drilling from American ships deep into the ocean floor. The Soviets proposed to contribute considerable scientific expertise to the project in addition to the money, expertise which had been developed during previous cooperative activities in related fields. The reason for the reversal was that the Soviets might gain access to advanced technologies with naval applications, and particularly seaboard navigation systems. However, there clearly was a political aspect of the decision. One year earlier, the same government agencies had approved the project. Also, the equipment which was in the center of the controversy was more than 10 years old, and Soviet specialists had already worked on American research ships with similar equipment.

In the same year the White House turned down another recommendation of the National Science Foundation to contribute $500,000 to the support of programs of the International Institute for Applied Systems Analysis in Vienna (IIASA)—an institution established by Soviet and American scientists as an East-West meeting place to address global issues of interest to the two countries. The stated reason for the denial was that direct bilateral cooperation was more effective than multilateral cooperation through the Vienna institute, which is now staffed by scientists from many countries. At the same time, top US

Government leaders had become convinced that the IIASA was a haven for Soviet spies with particular designs on Western computer technologies, and this view was a significant factor in the decision. In 1987 this decision was reversed as new personalities became involved in the reviews.

*   *   *

For many years the possibility of bilateral cooperation in science and technology was treated by the US Government as a bargaining chip in extracting concessions from the Soviet Government in other areas of higher interest to the United States. The US Government has always considered science and technology to be an area where the USSR lags considerably behind the West. Therefore, the US negotiating strategy was to hold back on cooperation in science and technology until the Soviets were more forthcoming in other fields, such as education and information exchanges. Now the US Government has apparently abandoned such linkages between scientific exchanges and other cooperative programs, although broad foreign policy considerations will inevitably play a role in reaching decisions on specific exchange activities.

If political relations deteriorate during the implementation of science exchanges, the traditional view has been that postponing the exchanges in science and technology is an effective way to apply pressure on the Soviet Government since such delays hurt the Soviets more than they hurt the United States. However, little evidence supports the contention that once agreements for cooperation are in place, politically inspired delays of exchanges can be used to influence Soviet policies in other areas.

Going back in history, we should recall that soon after the first scientific exchanges were carried out in the late 1950s, they earned the reputation as a particularly good trade-off in gaining

Soviet agreement for greater cultural contacts. For example, during the early 1960s the US Information Agency was having difficulty distributing in the USSR its magazine *Amerika*, which featured interesting snippets of life in the United States. This distribution was to be based on sales of the magazine at corner kiosks in various Soviet cities. The Moscow distributors of publications in the USSR and the kiosk vendors in the cities shared the lack of enthusiasm of Soviet officials for promoting *Amerika*. They were much more interested in distributing more politically acceptable publications in very large quantities.

The US Information Agency was also having difficulty with the Soviet authorities in arranging for large exhibits of American accomplishments in a variety of technical and cultural fields in cities outside Moscow. Such exhibits had earlier attracted throngs of curious Soviets in the smaller Soviet cities which had little exposure to the West.

Therefore, the Department of State delayed headway in arranging scientific exchanges waiting for progress in expanding the sales of *Amerika* and in arranging US exhibits. This tactic probably had some influence on Soviet attitudes in the information field, but it also reduced opportunities for exchanges of interest to the US scientific community.

During this same period, a high priority of the Soviets was the exchange of large artistic presentations, such as the Soviet ballet troupes and circuses, which could earn foreign currency during tours of the United States. While these and many other interests of both sides were eventually accommodated in the exchange arrangements which could then be hailed as balanced, much of the potential for scientific cooperation was lost in the process. Also, difficulties in implementing cultural exchange agreements took away some of the anticipated payoff from the exchanges. Defections of Soviet performers while on tour in the United States and demonstrations by anti-Communist groups before and during Soviet performances in some American cities resulted in Soviet retaliations, including cancellation of scheduled visits to the USSR by American artistic groups.

Policy officials from the United States have never fully appreciated the contributions of scientific exchanges in opening up Soviet society to Western ideas, which has always been a major US foreign policy objective. The lasting nature of international contacts among scientists and the willingness of scientists throughout the world, including within the USSR, to address issues logically and openly are consistently underestimated. Meanwhile, American officials repeatedly laud the potential impact of cultural exchanges on the behavior of Soviet artists and literary figures who can reach large Soviet audiences through their performances and writings.

As a group, the scientists of the USSR are more open to considering ideas from abroad than are many other segments of the society. In the glasnost era the Soviet media has become a free market for new and radical ideas; the Soviet scientific community has not hesitated to inundate this market with concepts obtained from the West—genuine competition among investigators for research funds, expanded research in universities, elimination of censorship of scientific articles, and greater opportunities for young scientists to travel to professional meetings. Thus, the concept of trading scientific exchanges for opportunities for greater contacts in other fields where there can be more political impact has always been very questionable. Hopefully, such thinking is behind us.

Meanwhile, Soviet science officials repeatedly protest the United States' approach of holding back their scientific exchanges until progress is made in other areas. They simply do not want to recognize reality; politics and science will always be linked to some extent with superpower relations. Soviet scientists will remain bored with the prolonged debates among officials of the two countries over the importance of "balanced" approaches. Their repeated arguments that if there is an important scientific problem which could benefit from cooperation, the two countries should address the problem without delay appeals to American scientists but will not carry the day.

*   *   *

Several examples reflect US efforts to tie very specific exchanges to foreign policy concerns over the years.

In 1965, the US Government abruptly cancelled a scheduled visit to the USSR by the world's only astronaut-aquanaut, US Navy Commander Scott Carpenter. The navy was upset over Soviet denial of a request for a US Navy oceanographic ship to call at Leningrad at about the same time as Carpenter's proposed visit. Oceanographic ships of other countries were being allowed to make the port call in connection with an oceanographic congress in the USSR, and US intelligence agencies were very eager to send a navy ship equipped with the latest electronic gadgetry into the naval port of Leningrad for the occasion. The Soviets were not interested in such sophisticated snooping in their own waters.

The Soviet authorities had gone to unusual lengths to provide Carpenter with access to a broad spectrum of influential political, media, and scientific leaders and Soviet cosmonauts. His visit was widely publicized in the USSR since at that time astronauts and cosmonauts were publicly revered for their accomplishments. Carpenter, who had lived at the bottom of the sea as well as in outer space, had very unusual and exciting credentials, with particular appeal in the USSR. Thus, by cancelling the visit and trying to punish the Soviets, the United States passed up this excellent opportunity for interactions of Americans with Soviet space scientists who seldom appeared in public. Also, the United States missed out on a public relations triumph.

In the early 1980s, in the wake of the Soviet invasion of Afghanistan, the United States backed out of a cooperative program of exploration of the White Sea on the northern periphery of the USSR after the Soviets had constructed a special guest house for American scientists in the coastal town of Chuba. In this instance an opportunity for a continuous American presence in a geographic area that is seldom open to Westerners was sacrificed for political reasons.

More recently, the Soviet Government made repeated overtures to the United States concerning their interest in cooperation in AIDS research. However, in 1985 and continuing through 1986, a number of articles appeared in the Soviet central press asserting that the AIDS virus was a by-product of US activities related to biological warfare carried out near Frederick, Maryland. The United States refused to cooperate in the field of AIDS in response to this disinformation campaign; indeed, cooperation would have been politically hypocritical.

Beginning in 1986 several eminent Soviet scientists spoke out in opposition to these false allegations about the United States, and eventually an impressive number of retractions were published in the Soviet press. Still, from time to time, Soviet articles concerning an alleged link between AIDS and biological warfare continued to appear in the provincial press. The Soviet authorities contended that they could not control yellow journalism in the age of glasnost, and the US Government steadfastly rejected overtures for cooperation in AIDS research. In 1987 and again in 1988, the new president of the Soviet Academy of Medical Sciences began to take a firm hand in helping set the record straight through signed newspaper articles. Finally, by mid-1988 the Soviet disinformation activities had ceased, and the first steps were taken by the two governments to cooperate in the field of AIDS research. The link between unacceptable political behavior and cooperative activities was very clear.

The travel plans of senior health officials of the United States to visit the USSR have been particularly vulnerable targets as the US Government displays its displeasure with Soviet behavior. During the past decade a number of planned trips to the USSR to discuss cooperation in the medical field have been postponed in response to Soviet activities: the use of psychiatry by internal security agencies to quell dissidents, the denial of adequate medical treatment for political prisoners, and most recently the disinformation on AIDS. Soviet policies have changed in all of these areas, and the American attitude toward cooperation probably contributed to these changes.

*   *   *

Since applications of science and technology provide much of the basis for the East-West military confrontation, cooperation in science and technology assumes a special symbolic importance in the relationship between the two superpowers. Kissinger was particularly successful in using scientific and technological exchanges to symbolize the favorable state of relations during and immediately following his tenure in office.

Many mementos—photos, pennants, statuettes—now adorn museums and research facilities in the USSR and the United States as tangible reminders of the spirit of détente of the 1970s. Soviet and American scientists who participated in these cooperative programs long for the good old days which are gradually returning. Also, a few of these scientists have discovered other routes through international organizations and private arrangements for carrying forward the cooperation that began through the formal intergovernmental exchange agreements.

At an early stage of space cooperation, a highly publicized direct data link was established between the US Weather Bureau and the Soviet Hydrometeorological Service for transmission of data acquired by weather satellites. While the Soviets could make good use of the data we were sending, we were receiving little of value from them. Their data were either too late or of too poor quality to be useful. However, as I traveled around the remote provinces of the USSR, I encountered numerous Soviet scientists who knew about the link and who used the American data regularly. Their gratitude was overwhelming and significantly affected their attitude toward the United States in general. Thus, I reasoned that the symbolism and the one-way exchange were clearly in our interest. I won my arguments at that time, although I had very few sympathetic supporters within the US Government other than the American ambassador in Moscow who strongly believed that the political benefits outweighed the lack of technical benefits from the program.

The political symbolism of scientific cooperation which reached its peak with the Kissinger agreements has been of limited importance within the United States. Cooperation in manned space flight, bringing together American and Soviet astronauts and cosmonauts, has been widely reported by the press and television and is generally considered to be a good way to save money while improving mutual understanding. However, few Americans are informed about or really interested in other aspects of US-USSR scientific cooperation. Of course visits by Soviet scientists to small American towns usually warrant front-page news coverage in the local press, and the *New York Times* dutifully reports the signing of each exchange agreement. As a result of this limited publicity, a few American scientists have become more interested in visiting the USSR and in engaging Soviet colleagues in more serious cooperation. This local enthusiasm has contributed to the general increase in people-to-people contacts between the two countries. But the overall political impact has not been great.

In contrast, the Soviets have always been eager to be seen as equal partners with the United States, and American scientists and scientific achievements were suddenly on center stage in the USSR as cooperation expanded during the 1970s. The Moscow press reported the signing of each agreement, with the reports systematically repeated in newspapers reaching villages across the country. These agreements were a clear signal that having contact with American scientists was acceptable, and the Soviet scientific community began to relax about interactions with Americans. The reports also imparted pride to the Soviet population, which for more than 50 years had been implored to catch up with the United States in science and technology. Now, Soviet scientists were working at the same level with their American counterparts.

In the USSR, as in the United States, cooperation in manned space flight was the most visible aspect of scientific cooperation during the 1970s. The constant Soviet propaganda on Soviet space accomplishments which permeated the press, television, and the movie theaters had a new angle: American achievements and cooperation were the order of the day. Exhibits recognizing this new element of cooperation opened at the main exhibit grounds and in the planetariums in Moscow and in the provincial capitals. Astronauts and cosmonauts rendezvousing in space became a theme that would last for more than a decade despite the swings in political relations. Lapel pins depicting the Apollo-Soyuz linkup in space were sold throughout the USSR.

One of the Kissinger-inspired agreements called for cooperation in new approaches to energy development and conservation. Specifically, the two countries decided to cooperate in the field of magnetohydrodynamics, a technique for improving efficiency in the generation of electricity. The US Department of

Energy, with great fanfare, sent a large magnet valued at about eight million dollars to Moscow, and the Soviets began to construct an experimental facility to house the magnet. This joint laboratory to test new technical approaches would indeed be a symbol of cooperation on the frontiers of science. When political relations began to deteriorate in the early 1980s, the United States withheld important components for the experiments; and the Soviets stopped construction of the special facility. A decade later the magnet sits in a corner of a research institute in Moscow, a symbol of the impact of politics on science.

For 25 years the presence of the Soviet fishing fleets off US coasts has been a major political issue. The United States has emphasized cooperation in biological research directed to preventing overfishing and to encouraging equitable divisions of fishing, as well as limitations on fish catches. Such cooperation often involves calls at US ports by Soviet fishing ships and exchange visits aboard ships by American and Soviet scientists conducting experiments to determine the sustainability of the fish populations. The political importance of this cooperation, particularly along the Alaskan coast, will continue to be very great.

In a related field, civilian oceanographic ships from the United States and the USSR occasionally make port calls in the other country. Given the swings in the political relationship between the two countries, these port calls rapidly become political events which command considerable local attention, particularly in the United States. The city fathers in Yalta and San Francisco, for example, usually take special pains to associate themselves with such visits of good will.

Foreign policy specialists in both countries usually take note of the state of exchanges in science and technology but generally consider such exchanges to be a minor part of the bilateral relationship. They have become accustomed to gauging the state of the relationship by the intensity of conflicts in the Third World, by progress on arms control, and by development of highly visible cultural contacts. Science and technology cooperation has been largely a minor affair, but an affair that received great political impetus during the 1970s and is currently being revitalized. Spacecraft, accelerators, research ships, and other types of scientific hardware simply are too photogenic to be completely ignored.

Despite the close intertwining of scientific cooperation and foreign policy, exchanges have at times been a misleading indicator of the state of relations between the two countries. For example, in 1964 and 1965 scientific exchanges were rapidly expanding, and I was regularly entertaining large delegations of American and Soviet scientists at my apartment within the US Embassy complex in Moscow. At the height of this euphoria about cooperation in science, demonstrators on the street below—who had been incited by the Soviet Government—stoned our apartment on three occasions in protest over US policies in Africa, Lebanon, and China. My wife and I had difficulty explaining to our young daughters why we were serving champagne to Soviet officials one day and sitting on the floor of our back bedroom the next as demonstrators pelted our living room windows with stones, with Soviet militia men standing idly by.

A final example in the field of environmental protection also shows that occasionally individual exchange activities are not necessarily dependent on the state of bilateral relations. During the early 1980s, when political relations were considered to be at a low ebb and bilateral science agreements were to be put on hold, more than 35 projects were progressing in a very quiet manner under the US-USSR environmental agreement. Ecological and health problems were simply too important to ignore, regardless of political difficulties. This activity was little known even within the Legislative Branch of the US Government.

*   *   *

Some US Government officials worry about the naiveté of American scientists who are dealing with Soviet colleagues. They consider many American scientists as liberals, sometimes at odds with the US Government over international policies. They fear these scientists will be easily taken in by Soviet propaganda and will be easy targets for Soviet intelligence agents masquerading as Soviet scientists. On the other hand, Soviet exchange scientists are perceived within US Government circles as having been carefully selected on the basis of their commitment to Soviet objectives at home and abroad and extensively briefed in Moscow as to Soviet intelligence interests.

Over the years a few American scientists have been seduced by Soviet prostitutes, entrapped by the Soviet intelligence network, and used to feed the Soviet propaganda machine. Fortunately, these cases are rare. Much more frequently, American scientists complain that Soviet scientific colleagues do not live up to their commitments concerning two-way exchange activities. Very often Americans provide Soviet specialists with reprints of their scientific articles, unpublished data, and other scientific materials expecting in exchange comparable information and materials which never appear. Soviet scientists encounter many practical problems in obtaining, reproducing, and mailing scientific materials abroad. Also, promises that never materialize are accepted as a way of life in the USSR, and this characteristic spills over to the international scene. Nevertheless, perceptions of the potential for political seduction of American scientists and the reality of one-way exchanges support American opponents of scientific exchanges who argue that our scientists will be too generous in sharing their innermost insights about high technology during interactions with Soviet colleagues.

Ever since the end of World War II, the antiwar drums have thundered loudly in the USSR. Many visitors to the USSR see one or more of the hundreds of memorials to the victims of World War II. I, and many other Americans, have been greatly moved when visiting the site of a former concentration camp in Latvia, walking among the graves of the defenders of Leningrad, or retracing the Soviet military breakout in Stalingrad (now Volgograd). The depth of concern among the Soviet citizens over the possibility that such a tragedy could be repeated is obvious even to the casual visitor to the USSR. During the Gorbachev era, prevention of nuclear war has become a more popular theme than ever before, and the Soviet scientific community has been handed the lead to spread this concern throughout the world.

Every year hundreds of discussions between Soviet specialists and American visitors to the USSR are devoted to the issues of war and peace. Leading Soviet scientists spend large portions of their time participating in these meetings. Concerns over war and peace also emerge during visits of foreigners to Soviet villages, factories, and farms. It seems that the most important message that every Soviet wants to convey to American visitors is that war must be avoided.

In Moscow, large convocations of scientists from around the world to address important international issues have become regular events during the last several years. These meetings usually include one or more antiwar themes. The themes have included the medical consequences of nuclear war, disarmament, and activities in outer space. Attendance by large numbers of prominent American scientists is considered by the Soviet organizers to be critical to the success of the meetings.

The Soviets usually cover all expenses for hundreds of foreign invitees, including international air travel via Aeroflot.

Many American scientists flock to these events. Years ago, some Americans had been hesitant to accept free tickets and free hotel rooms, lest they be perceived as Soviet collaborators. This hesitancy has now faded as the scientists have convinced themselves that they are contributing to important peace-oriented activities. US Government officials are often among the attendees, which sends a clear signal that participation is fully acceptable.

The heavy media coverage by the Soviet and foreign press tends to turn these events into glamor gatherings. Nevertheless, during the past several years the foreign visitors have been very pleased with the substantive aspects of their participation. The programs have been masterfully orchestrated by the Soviet hosts to maximize opportunities for meaningful discussions, for meetings with the highest leaders of the USSR, and for visits to a wide range of institutions which have adopted an open-door policy. Gorbachev personally participates, and he is a master at mingling with visitors.

*   *   *

For many years the American ambassadors in Moscow have argued that a very important benefit from exchanges is visitors' impressions of the culture of their host's country. There is no doubt that Khrushchev's visit to an Iowa farm had a tremendous impact on his views on agricultural policies and that Reagan's walk through Red Square with Gorbachev will long influence his views on the USSR. At a lower level, trip reports of Soviet and American exchange scientists are replete with commentaries on the impacts of the visits on their perceptions of life in the other country; in the United States these reports are usually sent to interested officials and scholars.

Many of the current political and scientific leaders of the two countries have visited the other country earlier in their careers. Though they made subsequent visits when they reached positions of higher authority, many say that the early visits provided the most lasting impressions in their minds and that the later visits reinforced these impressions. What is the nature of the impressions? Do they help or hinder the process of improving political understanding between the two countries?

As we will see in the next chapter, in a few cases exchange visits disappoint the participants. Administrative problems with the travel, lodging, or programmatic arrangements sometimes sour the visits. Occasionally, visitors are victims of lost items or theft. Some exchangees believe they were denied access to facilities or people for political reasons, and the associated bitterness detracts from the other aspects of the visits. American visitors to the USSR during the summer simply don't believe that Soviet colleagues whom they thought they would see leave home for six weeks' vacation every year; but in reality Soviets seldom forgo their lengthy vacations. Soviet visitors to the United States cannot understand why they can go to some universities while the American visa authorities deny them access to other institutions which have also extended invitations to them.

Most scientific exchangees from both countries, however, consider the experience a highlight of their lives. They have combined a political adventure with scientific inquiry within a fascinating cultural milieu. Among the most enthusiastic visitors are those who have experienced small town hospitality during their visits—a more common occurrence in the United States than in the USSR, where good science is more highly concentrated in a few major cities.

Overall, exchanges clearly enhance the tolerance of the participants for the differences in the political systems of the two countries. If there are tensions between the two countries, the exchange scientists blame the governments, not the common people.

*   *   *

Foreign policy shapes and constrains scientific cooperation. Formulation of foreign policy also greatly benefits from information developed during exchanges, and particularly exchanges in science and technology.

Traditionally, the US Government has relied on reporting from our embassies abroad and on information generated by the intelligence agencies as the basis for confirming or modifying our foreign policies. In the case of a country as large, as complex, and as secretive as the USSR, there have always been severe limitations on the adequacy and reliability of information obtained through these channels. Now in the age of glasnost floods of information are coming forth, and a serious problem of filtering out the reliable information exists.

Scientific exchanges will continue to play a very important role in providing authoritative insights of direct relevance to US foreign policy decisions. In the 1960s and 1970s exchanges in the space and nuclear sciences, in particular, were critical in ensuring a good appreciation of Soviet capabilities for the foreseeable future as we developed our policies in related fields. Much of our information about Soviet computer capability and Soviet advances in biotechnology can also be traced to exchange activities.

Exchanges should not be co-opted for more effective intelligence collection. Exchanges can provide, as a by-product, interesting intelligence information—often information of a totally unanticipated nature. Still, exchanges have a far broader purpose and a much longer time frame in promoting improved mutual understanding of important scientific issues and of the cultures of the two countries.

*   *   *

Both the United States and the USSR expect scientific cooperation to serve many political purposes in addition to the advancement of science. The many motivations of the past for promoting or limiting bilateral exchanges will undoubtedly continue to characterize Soviet and American attitudes for the foreseeable future. But the basic justification for exchanges should remain the scientific benefits that are derived, benefits that are directly dependent on the stability of long-term cooperative relationships.

The scientific importance of long-term cooperation on global problems is clear. The future payoffs from friendships that are established by young scientists early in their careers are gradually receiving greater recognition in both countries. However, the short-term gains and losses from exchanges remain a dominant factor in the political decisions to move forward with specific exchange activities.

As a practical matter, scientific exchanges will continue to be turned off as one country develops policies that conflict with important interests of the other country. They will be turned on as summits approach and as new leaders seek to gain recognition as peacemakers.

Scientists will simply have to live with these realities. They need to design their cooperative programs to the extent possible in modules, with each module achievable in a relatively short time. This is not easy since research is a continuum of trial and error and depends on small but steady incremental advances. If long-term experiments are critical to advances in a specific field, perhaps research in that field should not be pursued on a cooperative basis. After the inevitable pauses in cooperation, succeeding modules may be able to build on earlier efforts without too much loss of momentum due to the exigencies of foreign policy.[1]

CHAPTER 6

Scientific Cooperation on the Rebound

What is national is no longer science.
Anton Chekov

The growing stream of American and Soviet scientists being processed through the Pan Am and Aeroflot counters attests to the recent upsurge in the scope of scientific cooperation between the United States and the USSR. While scientists inevitably spend an enormous amount of energy coping with the administrative problems and cultural adjustments before and during their exchange visits, they nevertheless find considerable reserve energy within themselves to carry through. Sleep simply assumes a lower priority than usual, and the increased flow of adrenalin helps most exchange scientists take advantage of professional opportunities during their visits.

Scientific cooperation between the United States and the Soviet Union encompasses a very wide range of activities. This cooperation draws on American scientists from many types of institutions—from the government, from industry, and from the academic community. Similarly, Soviet scientists from many institutions participate as exchange scientists, as hosts for visiting Americans, and as correspondents with American colleagues working on related scientific problems. Their interests range from studying the galaxies to investigating the polar ice caps, to comparing geological rift systems of different continents, to tracing the origins of acid rain, to enhancing energy conservation efforts, to combating the spread of viruses, to cataloging butterflies along the Alaska-Siberia frontier.

During the past several years, bilateral scientific exchanges between the United States and the USSR have again been on the rise. Over the years the expansions and contractions of scientific exchanges have mirrored the trend in exchanges in general, including exchanges of students, artists, historians, and language specialists. As previously discussed, this trend has been very sensitive to the state of US-USSR political relations—a steady increase during the détente of the 1970s, a sharp decline following the Soviet takeover of Afghanistan in 1979, and rapid growth after the first of the Reagan-Gorbachev summits in 1985 and continuing to the present.

*   *   *

A relatively new aspect of cooperation is the many arrangements between American and Soviet commercial organizations that encompass some aspect of science or technology. Cooperation in science is usually designed to lead eventually to sales of products or licenses. However, as noted below, a variety of administrative and financial barriers inhibit moving from modest cooperation in research and feasibility studies to joint ventures and other ambitious commercial schemes.

A common belief within the American business community is that many Soviet citizens with large ruble savings would spend these savings instantly if they could use them to buy high-quality consumer goods produced in the West. Many firms from around the world would like to provide these goods. However, they want hard currency in exchange: rubles aren't worth very much in New York or Tokyo. The official exchange rate is $1.60 for one ruble; but the real value of the ruble is better reflected in the black market rate of five rubles for one dollar on the street outside any Leningrad hotel. The Soviets are reluctant to use their foreign exchange earnings from exports of oil, sable, and gold to pay for imports of goods that they feel they can produce locally. Nevertheless, many foreign firms believe that they can work out barter or other types of arrangements that will enable them to eventually recover their costs of doing business in the USSR and to earn an adequate profit in currency which can be used in the West. Also, these firms would like to be entrenched in the USSR should the ruble someday become internationally convertible. According to Soviet colleagues, in November 1988 the Politburo of the party set 1992 as the target date for a convertible ruble; given the historical isolation of Soviet financial institutions from international markets and the currently depressed state of the Soviet economy, this target seems unrealistic.

The US-USSR Trade and Economic Council was established a decade ago to help coordinate the activities of more than 400 member firms from the United States which have been actively exploring commercial opportunities in the USSR. The Council maintains offices in Moscow and New York that are staffed by both American and Soviet specialists. It provides information for the firms and for others interested in economic developments in the USSR. It sponsors dialogues between Soviet political and trade leaders and American officials and business executives and generally supports intergovernmental programs related to trade.

About 100 commercial arrangements now exist between Soviet and American organizations. In some cases, they include joint research projects being carried out in the United States and the USSR. They also include exhibits, joint seminars on technical topics, advisory services, visits by specialists to explore specific technical areas, and sharing of technical data. They involve, for example, sensitizing the Soviet taste to the ever-secret formulas for Pepsi-Cola, Coca-Cola, and Fanta; introducing automation systems into Soviet fertilizer plants; improving the quality of nutritional supplements for animal feeds in the United States and the USSR; applying unique Soviet medical instrumentation in American institutions; marketing Soviet agricultural equipment in Wisconsin; and jointly publishing magazines on recent developments related to personal computers.

Within the USSR, many research institutions and leading specialists participate in the joint endeavors. Unfortunately, new bureaucracies at many levels have been created in Moscow simply to serve as intermediaries with the Western firms despite the Soviet policy to cleanse the system of unnecessary bureaucrats. Also, as the Soviet Government tries to decentralize the authority for entering into international commercial arrangements from the ministries to the industrial enterprises, the enterprises which do not have international experience hesitate to make decisions. Doing business in the USSR is not easy.

Many American scientists and engineers visit the USSR to support existing arrangements and to explore opportunities for making new deals. Some companies have offices in Moscow, usually staffed by one or two Soviet employees. A large number of sales representatives of US firms have also logged many tens of thousands of miles in never-ending efforts to keep agreements on track.

While traveling in Central Asia last year, I met a West European sales representative of an American firm who has spent about six months each year for the last 15 years in the USSR selling scientific instrumentation to laboratories all over the country and establishing the supporting maintenance network. He has learned not to rely on intermediaries but to make all arrangements himself, from obtaining the commitments by the ministers for his products to gaining the support of the individual researchers who use the instruments. His persistent efforts have paid off. His firm has placed more than 400 instruments throughout the country, with a sales value exceeding $10,000,000. In every case he has received prompt and full payment in hard currency, largely because the researchers know that the instruments will work and will be repaired promptly if necessary; with such assurances, the ministries have been willing to divert scarce foreign currency for this effort.[1]

*   *   *

For more than 25 years a number of US Government agencies have had formal agreements with counterpart agencies in the USSR involving exchanges in science and technology. The National Oceanographic and Atmospheric Administration wants to share in the fruits of the Soviet's massive collection of oceanographic data. The Department of Housing and Urban Development is interested in Soviet experience in construction in Arctic regions. The National Park Service hopes to learn from Soviet research in its extensive network of preserved nature parks. The Department of Energy and the Nuclear Regulatory Commission do not want to see another Chernobyl-type accident anywhere in the world and are prepared to help the USSR improve its nuclear safety programs. The Geological Survey studies volcano systems throughout the world. The National Institutes of Health plan to learn from Soviet experiences in coping with alcoholism.

The agreement between the two governments in the environmental field has been a pacesetter for bilateral cooperation in recent years, involving several dozen projects directed to environmental planning, natural resource management, environmental pollution, and related areas. Soviet officials and scientists are belatedly awakening to the very severe unattended environmental problems in the USSR—the pollution of Lake Baykal, the drying up of the Aral Sea, the brown clouds over the Siberian cities, the discharge of toxic wastes into waterways. They are interested in learning from experiences in addressing similar problems in the United States. While the United States is ahead in most aspects of controlling pollution, cooperation has benefited American specialists in the development of new environmental assessment techniques, in relating pollution levels to ecological damage, and in broadening gene pools to help preserve rare plant species. Also, global environmental problems have become a focus of a great upsurge in scientific interest in both countries. The USSR occupies one-sixth of the earth's land surface and must be a major participant in any serious international effort to address global environmental issues. Sulfur deposition in the Ukraine attributable to the power plants of Central Europe, the warming of the biosphere which could result in Arctic snowmelt, the pollution of the world's oceans which wash the coastlines of the USSR, and the worldwide acidification of soil which impacts the huge Soviet landmass exemplify some of the current Soviet concerns.

In addition to the inherent importance of these topics, cooperation in addressing very broad issues such as the effect of pollution on the global climate lets Soviet scientists display their very sophisticated talents in the field of mathematics, and particularly in the development of theoretical models to simulate conditions over large polluted areas. At the same time, Soviet scientists are not required to dwell on the microproblems of environmental degradation, such as air pollution in the Siberian city of Bratsk, for example; seldom can they influence abatement policies which are in the hands of the production ministries, and often they are not well informed about the problems.

The most memorable bilateral agreements have been in the field of space exploration. Cooperation, as noted earlier, began during the 1960s as both countries dramatically expanded their manned and unmanned space programs. Cooperation in space medicine resulted in genuine sharing of information on survivability in space. Early exchanges of close-up photographs of the moon were also important in unraveling the history of the solar system.

Several personal experiences are related to space cooperation. At four o'clock on a Sunday morning in 1964 while I was serving as the Science Officer in the American Embassy in Moscow, I was summoned by the duty officer to the communications center in the Embassy to handle a "flash" cable from Washington. The cable reported that Soviet radio signals were interfering with the flight of our manned spacecraft Gemini 4 as it passed over the Western Pacific and over the territory of the USSR. "Get the Soviets off the radio frequency we are using. They're endangering the lives of our astronauts," read the cable. NASA had not announced to the Soviets or to anyone else the communications frequencies being used by the spacecraft, since they were worried that if the frequencies were not kept secret, amateur radio operators might learn about them and try to interfere with the flight.

Arousing Soviet officials at their weekend houses on the edge of Moscow at that hour and persuading them to instruct Soviet military commanders to turn off their electronic devices on Soviet ships and at surveillance stations on the Chinese border seemed like an impossible task. Fortunately, many months earlier I had sequestered in my night stand several emergency telephone numbers at the Academy of Sciences, the Ministry of Foreign Affairs, and the Ministry of Communications. By allowing the phone to ring for more than five minutes I was able to rouse a groggy night watchman at the Academy. At the time, there were no greater heroes in the USSR than the cosmonauts. Thus, when I shouted, "You are threatening the life of an astronaut," his response was immediate. He would call the president of the Academy, the vice president, the scientific secretary, the president's chauffeur, and anyone else he could find. And he did.

For the next two hours the telephone in our apartment rang off the hook, while my sleepy family groaned in the background. With the Academy of Sciences, the Ministry of Foreign Affairs, the Ministry of Communications, the State Department, NASA, and the ambassador all calling, I felt like the most popular person on the globe. By six o'clock the Soviet transmitters had changed frequencies, and I went back to bed. This was scientific cooperation in action, at least from my vantage point.

The following year I became involved in another type of space cooperation. I saw a stirring Soviet television broadcast of the first walk in space by Soviet cosmonaut Leonov. Our astronaut Ed White, whom I had known when we were cadets at West Point, was scheduled to make a similar walk a few weeks later. The next day while I played tennis with Frank Bourgholtzer, a well-known American NBC correspondent in Moscow, we talked about the film and his possible interest in showing it on American television. We agreed that it would have wide public appeal and might also help Ed White prepare for his walk. Frank subsequently purchased a copy of the film from a Soviet news agency, and it was promptly shown to the American public and to Ed. Three months later, Ed bought me lunch in Houston and described how the film had helped him in learning to tumble in space without losing control.

By 1988 the two nations had come a long way in space cooperation. Astronauts and cosmonauts had met in space, exchange visits to launch sites had become routine, American instruments had flown on Soviet spacecraft, and many books had been jointly written by Soviet and American space scientists. Still, I was impressed by the changes that American colleagues encountered at the Space Research Institute in Moscow during several meetings in 1987 and 1988. The discussions were open and freewheeling and the exchange of information uninhibited. Twenty-five years earlier, I could not have even found the institute where space research was carried out in Moscow.[2]

The success of most intergovernmental agreements hinges on active participation by US Government laboratories and working scientists from the laboratories. However, a high percentage of the US participants in the programs are political appointees—though scientists—who are removed from research laboratories but who are eager to see the USSR firsthand and to use their negotiating skills. They travel to the Soviet Union, receive Soviet visitors in the United States, and then pass on to other pursuits. Consequently, considerable effort is expended by the civil service scientists in educating successive generations of political leaders in order to preserve the relationships that have been developed with Soviet counterparts. In this era of glasnost and perestroika the Soviet participants are changing just as rapidly. The fact that any progress is made is often remarkable.

There are seldom special funds available to US agencies to support their cooperative programs with the USSR. The participating agencies must divert funds appropriated by the Congress to support domestic programs, arguing that the United States will indeed benefit from cooperation. The agencies make this argument with considerable vigor, and to some extent the argument is sound. While serving as the director of the EPA laboratory in Las Vegas in the early 1980s, however, I was occasionally annoyed when money was taken from our research budget to pay hotel bills for visiting Chinese and Soviet scientists who were guests of EPA, even though I was obviously sympathetic to these types of exchanges. Some of my colleagues at other laboratories which were also losing their budgets were more than annoyed; they were furious.

Still, the opportunity to participate in cooperative endeavors with the Soviets holds a special fascination for most US scientists who control funds for research. A trip to the USSR or the chance to host a cocktail party for Soviet scientists in the suburbs of Washington or a Western barbecue in Nevada offers an opportunity to participate in diplomacy.

However, international science programs should not be placed in direct competition for funding with national programs which are judged on the basis of near-term technical benefits to the United States. The payoffs from international science are longer term, are more difficult to predict, and extend far beyond simple transfers of technical information.

                              *   *   *

All scientific institutions in the USSR belong to the government. Therefore, from the viewpoint of the Soviets, all international scientific cooperation has an official character. In the United States, however, a wide variety of private organizations collaborate with Soviet scientific institutions. Many of these American organizations have difficulty understanding that all Soviet scientists work for the state and that the concept of a research scientist who sets his own agenda is foreign to the Soviet lexicon.

In deference to American thinking, the Soviets have now begun referring to exchanges involving American universities and other private institutions in the United States as private exchanges, even though both sides are aware that the governments play a role, particularly on the Soviet side. These exchanges have varying degrees of formal endorsement from the political bodies of the two governments. Such endorsement ranges from explicit recognition of such exchanges in intergovernmental agreements to simple issuance of the visas necessary for the cooperation to move forward.

As one example of the nature of a private exchange, more than 600 scientists from the United States and 600 from the USSR have participated in individual exchange visits ranging from one to twelve months, sponsored by the National Academy of Sciences and the Academy of Sciences of the USSR. The participants have conducted research in many laboratories throughout the United States and the USSR. They have traveled from Alaska, to Hawaii, to Maine, and from Riga on the Baltic coast to Yerevan on the Turkish border, to Magadan and Vladivostok on the eastern coast of Siberia.

Also, about 25 bilateral scientific workshops have been held under the interacademy program—in physics, mathematics, biology, chemistry, earth sciences, and other disciplines. These workshops usually bring together 10 to 20 Soviet scientists and a comparable number of Americans for several days of intensive discussions of recent developments in the specific fields of interest. The program also includes visits to interesting research laboratories in the host country.

During the past several years, this interacademy program has expanded significantly. Many cooperative activities in addition to the exchange of individual scientists and scientific workshops are under way. Seismologists are investigating indicators of future earthquakes in California and Armenia. Geologists are comparing the origins of formations in the Caucasus and the Sierras. Energy specialists are seeking new ways to reduce the demand for limited energy resources. Environmental scientists are investigating the warming of the globe. Social scientists are constructing computer models of the causes of international conflicts. And leading scientists from both countries meet regularly to consider the technical aspects of arms control and international security.

The program of the two academies is the largest of the cooperative scientific programs conducted outside formal intergovernmental channels. It has been of special importance during the times of political difficulties between the two countries. As a nongovernmental activity, it has been somewhat insulated from political pressures and has continued at times when other programs have been stopped for political reasons.

Many American universities and professional societies have long been interested in Russian language training and in research in the social sciences and humanities in the USSR. Centralized exchange programs to facilitate cooperation in these fields have been in place for many years, and a large number of former participants in these programs now hold key positions in the public and private sectors in the United States and within the Soviet Government. The Soviet authorities often use these exchange programs to place Soviet scientists and engineers at US institutions, primarily as advanced graduate students, as the quid pro quo for arranging placement for American students and researchers in the social sciences and humanities in the USSR. This apparent asymmetry of interests has led American opponents of exchanges to contend that the United States sends poets to rustic Russian villages and the Soviets send engineers to Silicon Valley.

The Fulbright program has for several decades supported exchanges of scientists and scholars between the United States and many countries. It continues to be an effective mechanism for enabling American and Soviet scientists to spend academic years abroad in the USSR and in the United States. As is the tradition with this program throughout the world, the quality of the participants is generally high.

At present, more than one dozen American universities have direct links with Soviet universities calling for exchanges in many fields, including science and engineering. Often such linkages enrich sister city programs. Indeed, university-to-university arrangements are a very popular way to give tangible meaning to the sister city concept. Every politician—American or Soviet—likes to identify with the university in his or her town.

More scientific exchanges are now taking place on an informal basis than ever before. A recent upsurge in the numbers of invitations for visits being extended by scientists and institutions on both sides is quite apparent. Invitations to Americans are frequently linked to large meetings in Moscow or Leningrad when the Soviets consider the American presence important, such as meetings on Arctic research or the health effects of the Chernobyl accident. The opportunities for American scientists to have interesting scientific tours and discussions in conjunction with such large affairs are usually extensive.

Also, the positive responses of Soviet scientists to invitations from individual American scientists and institutions to visit the United States have been increasing. Past problems in arranging for specific Soviet scientists to attend scientific meetings in the United States are legendary; for many years last-minute cancellations by invited Soviet speakers, panelists, and commentators have been commonplace. Now the outlook seems much brighter.

Overall, however, the quantity of bilateral scientific interactions remains very low in comparison with the numerous interactions of American scientists with colleagues in other countries with significantly less scientific capability than the USSR. While there have been many discussions between Soviet and Ameri-

can exchange organizations about decentralizing the management of scientific exchanges as a step to increasing these interchanges, exchanges are still very much influenced by the central
political bodies of the two governments.

*   *   *

Numerous testimonials document the scientific benefits of
bilateral cooperation—to the East, to the West, and to the international scientific community. Hundreds of American and Soviet participants in exchanges have reported their experiences in a
wide variety of scientific and technical journals and books, frequently as joint publications of Soviet and American authors.
Also, they have written articles for the lay audience in their
countries, given public lectures, and made presentations at scientific conferences based on their experiences as exchange scientists. Films of some exchanges have been adapted for television and movie audiences.

In the commercial arena, the best testimonials are advances
in industrial and agricultural processes and products which can
be traced in part to cooperative programs. While there have
been some examples of direct commercial benefits from cooperation in science and technology, to date the return for most
American companies has not been commensurate with the investment of time by corporate leaders and scientists. For example, of the 400 senior executives who visited Moscow in the
spring of 1988 in search of commercial opportunities, only a few
can show concrete results from the visit, and the financial payoff
from most of these undertakings is uncertain. Nevertheless, the
American business community gives considerable importance to
keeping abreast of developments in a country as large as the
USSR. Cooperation in science and technology provides a good
window for assessing the future commercial potential of dealings with the USSR.

The US Congress has conducted periodic hearings that have helped clarify the benefits and difficulties in cooperation with the USSR in science and technology. However, those who participate in these hearings find only one or two congressmen in attendance and a light sprinkling of specialists from the government agencies. While many members of Congress are interested in Soviet-American relations and many are deeply involved in the development of the scientific base of the United States, few have given a combination of these two concerns high priority.

More than 25 reviews of different cooperative programs in science and technology have been carried out by US Government agencies during the past decade, but the distribution of these reviews has been very limited. Unfortunately, many key policy officials have their minds made up concerning the costs and benefits of exchanges, and they don't want to take the time to seriously examine the record. In most reviews, the conclusion has been that there can be important scientific benefits to the United States as well as the the USSR from carefully designed and properly managed exchange programs.[3]

The US Government agencies that arrange cooperative programs prepare extensive budget justifications for their activities. They cite US reliance on the Soviet-designed Tokamak device in the development of fusion research in the United States. They point out the many lessons learned from Soviet space probes of the planets. They note the contributions of Soviet glaciologists to improved understanding of the environments at both the North and South Poles. They also caution about the pitfalls in cooperation and identify areas in which cooperation has not paid off, such as in the development of photovoltaic cells for solar power. Overall, the international affairs officers in the agencies generally believe that cooperation with the USSR is quite useful. Every trip to the USSR provides unexpected insights, and maintaining close contact with a large country of surprises is very important to them. While American scientists may not always gain new technical knowledge that can be applied today, they should be ready to take advantage of new Soviet approaches of the future.

Examples of cooperative activities that have been very beneficial to American working scientists can be seen in almost every important American scientific journal. A Soviet visitor to Harvard contributed to the scientific research that led to a Nobel Prize for an American specialist in DNA sequencing. Soviet mathematicians in Leningrad helped a visiting American from Purdue solve an important mathematical theorem that had eluded mathematicians for 30 years. An American visitor to Soviet freshwater fishery laboratories gained knowledge that was helpful in restocking Wisconsin lakes.[4]

It is often difficult to attribute specific scientific advances to a particular time period in a scientist's career. Advances usually result from an accumulation of knowledge and experience. International exchanges simply add a new dimension to this educational process. They may suggest new research methodologies or discredit old ones. They may provide new data that support or contradict existing data. They may uncover new scientific uncertainties that require attention. In an event, in a significant number of cases the exchange activities of American and Soviet scientists certainly accelerated the timetables for opening new scientific horizons.

*   *   *

Exchange programs have provided American scientists with important access to previously inaccessible geographical areas, research facilities, and data sources in the USSR. In a country as large as the USSR, such access is very important in many fields of science—geology, geophysics, zoology, botany, oceanography, coastal morphology, and atmospheric sciences.

Traveling outside the major cities of the USSR is not easy for Soviet citizens and often impossible for foreigners. Special permits are frequently required, usually for internal security reasons. Accommodations that are considered suitable for visitors may not exist. Transportation may be difficult. Finally, there may not be precedents for travel to interesting areas, and the sluggish administrative systems in the USSR can easily thwart attempts to do something new.

Scientific exchange programs have on a number of occasions provided the vehicle for breaking through these types of administrative barriers and for encouraging special arrangements by the Soviet authorities to receive foreigners in scientifically important but seldom visited areas. American ecologists have climbed the rugged mountains of the Caucasus and Central Asia. Wildlife biologists have traveled into isolated areas of central Siberia. Arctic specialists have conducted research along the northern coast of the USSR. American scientists have traveled by horse, by small boat, and by helicopter, sometimes introducing their Soviet colleagues to new geographic areas.

Access to Soviet laboratories is also important for American scientists. Such visits are often the only way to have contact with many young Soviet scientists. These scientists are increasingly uninhibited when receiving American colleagues in their institutes, and discussions are candid and detailed. Only by seeing experimental facilities can American scientists make reasonable judgments as to the quality of research and the significance of findings that are reported by Soviet investigators. Also, the system for publishing scientific data and results in the USSR is not well developed. Thus, discussions with Soviet bench scientists provide opportunities to gain access to data and results that might be delayed for several years before reaching the West or that might never reach the West at all.

*   *   *

Bilateral exchange activities are not uniformly productive, however. Administrative arrangements, particularly in the USSR, can be so poor as to negate the potential scientific value from an exchange visit. Cooling heels while waiting in a hotel room for the telephone to ring with news about appointments which have been requested is indeed frustrating. Arriving for serious research at an institute that is unaware of your visit until the last minute discourages even the most flexible scientist. Learning that key colleagues are traveling elsewhere during the time of a scheduled visit reduces enthusiasm for effective inter-actions. Of course these types of problems are also encountered in the United States; but our formal and informal communication systems are much better developed, and the administrative problems for Soviet visitors are much less frequent.

Not all topics lend themselves to exchanges. In the past, poor selection of topics has sometimes resulted in mismatches of interests and capabilities. For example, in 1988 I accompanied to the USSR an American delegation of scientists interested in new approaches to the development of vaccines. The Americans were disappointed to learn that many Soviet research programs were still in their early stages, and a visit in several years proba-bly will be more profitable. I also traveled with Soviet energy specialists to the wind farms east of San Francisco in 1987, but the Soviets saw little of relevance to their interests in American attempts to use wind power for generating electricity; for them, wind power is not an economically attractive energy alternative in the USSR.

Tourism sometimes becomes a principal motivation for American scientists to participate in exchanges. Sightseeing takes precedence over scientific discussions on occasion. The ballet and the monastery visits can set the timetables. Not that these cultural experiences have no value; it is simply a matter of priorities in programs that are always too short in duration.

Meanwhile, some US Government officials press for exchange activities to take place only in the USSR, since they consider exchanges in the United States to be of less interest. In some cases, this bias reflects the influence of US intelligence agencies whose primary task is to gain information on scientific activities within Soviet facilities while limiting Soviet access to US facilities. The bias also reflects a concern that US scientists will emphasize "show and tell" for Soviet visitors rather than mutually beneficial dialogues.

However, some exchange activities should be carried out in the United States. Many visits to the United States by Soviet scientists have been highly rewarding for both countries in the past. Indeed, the quality of Soviet scientists visiting the United States is often very impressive, particularly in recent years, and their scientific productivity very high. Serious collaborative efforts in the USSR are frequently very difficult. Given the underdeveloped state of the Soviet scientific infrastructure (space, equipment, spare parts, services, supplies), effectively including an American scientist in the research system there is much more difficult than providing scientific research opportunities for visitors to American laboratories.

*   *   *

For many years the USSR has promoted scientific exchanges with countries throughout the world. The Academy of Sciences of the USSR, for example, now has formal agreements with counterpart institutions in more than 65 countries.

The population of foreign scientists working in Soviet research institutes is quite large, often exceeding several dozen in a single institute, with most of the visitors from Eastern Europe, Cuba, and Vietnam. These visitors from the Socialist countries usually remain in the USSR for one to three years working on advanced degrees in science. The Soviet scientists draw on these visitors as research assistants but generally do not expect them to make major contributions to research activities. By contrast, relatively few visitors from the West are encountered in the institutes, and particularly in institutes outside Moscow and Leningrad. Most of the Western visitors are senior scientists on short-term visits of one to two weeks.

Similarly, we frequently meet Soviet scientists in the research institutes of Eastern Europe, and many reports indicate that Soviet scientists are carrying out research in Cuba and in Southeast Asia. The Soviet visitors are generally good, mid-level scientists with sound research projects. The host institutes for Soviets in Eastern Europe are usually the best research institutes of the region with research profiles close to the profiles of Soviet institutes. For example, there are close ties between the catalysis institute in Novosibirsk and the catalysis institute in Krakow, between the physics institute in Sofia and the physics institute in Dubna outside Moscow, and between the geography institute in Bucharest and the geography institute in Moscow.

As a matter of policy, Soviet scientific leaders assert that since science is universal in its premises and in many of its applications, international cooperation with many nations must be a key aspect of the Soviet scientific effort. The Soviet scientific approach to international cooperation is highly centralized and highly sensitive to political interests. Indeed, agreements for cooperation in science and technology frequently call for more projects than are possible simply to serve a political objective. Leading Soviet scientists often complain about how they have been overcommitted to help fulfill ambitious agreements signed by political leaders. While many countries are flattered by the high level of the Soviet scientists who initially participate in cooperative programs, this level of participation rapidly declines as these scientists must direct their international activities toward other countries. The intensity of cooperation may also decline in the absence of the senior Soviet participants.

Scientific cooperation with the United States has always enjoyed a special status within the Soviet scientific community. Soviet scientific leaders assign high priority to interactions with American colleagues at home and abroad. Now as Gorbachev reaches out to the West on many fronts, expanded scientific cooperation with the United States is viewed within the USSR as of great importance as a short- and long-term investment of both financial and manpower resources. During the past several years, the increase in Soviet awareness of the international dimensions of energy, health, and environmental issues has paralleled the heightening of similar concerns in the United States and elsewhere. Thus, the Soviet scientific community displays a continued readiness to cooperate with American colleagues on many fronts to address these types of critical issues.

Soviet scientists take great pride in their publications in Western journals, publications which are often facilitated through collaborative efforts with Western colleagues. In some cases, the only route for Soviets to publish is participation in exchange programs. Publication in Soviet journals can be a tedious process due to the review procedures, the unilateral control of many Soviet editors as to which articles should be published, and the priority given to senior scientists within a publication system that is always overloaded with proposed articles.

A scientific trip to the United States is highly prized. The opportunity to travel is often used as a "perk" for scientists who have made contributions to their institutions in the political arena as well as in science. While trips to other countries are always welcomed by Soviet scientists, visits to the United States have a special attraction in most fields of science.

Election to membership in international scientific organizations and in Academies of Sciences of Western countries is also a great honor for Soviet scientists. The recognition attendant to such election has on occasion been decisive for Soviet scientists in gaining scientific promotions within the USSR. At many levels the Soviets seek wide recognition for their accomplishments in science and technology. Simply sitting at the table with American specialists is an important form of such recognition.

Soviet officials and scientists also attach considerable importance to the cultural experience associated with visits to the United States, and particularly the opportunities to see the American countryside and to practice the English language. Soviet officials do not seem concerned when a heavy element of cultural activity limits the time available for scientific pursuits.

*   *   *

Participation in exchange programs can be both exhilarating and frustrating for the American and Soviet scientists. The experiences are unique; they simply are unparalleled by other types of international interactions among scientists. Thus, in assessing the scientific and political payoff from exchanges, the two governments must take into account the practical dimensions of carrying out the exchanges.

Arriving in Moscow is always a traumatic experience for American exchange scientists. American scientists invariably ask themselves: Have my Soviet hosts really been informed I will be on this flight? Will they find me at the airport? If they aren't here, I don't have the name of the hotel. I don't have any rubles. I don't speak Russian. Did my baggage make the connection in London? What will I do if it didn't? Will the customs officials really go through all of my baggage? What will they do with those American magazines I have brought along?

Most of the time, after a delay of one hour or more, the Soviet hosts find the visitors in the airport. The baggage arrives, and custom formalities are completed without incident. But difficulties do sometimes arise. Almost every day American scientists who are not met on arrival due to administrative confusion plead with Intourist representatives to take care of them for one night until they can find their hosts the next day. Travelers besiege Pan Am, trying to trace lost baggage. For unexplainable reasons, flights to Moscow are plagued with an unusually high rate of misplaced baggage. Since rubles cannot be brought into the country legally, the lines at the bank at the airport are long as foreigners from many countries seek rubles to pay taxi drivers for transportation to downtown locations.

Even the best accommodations for exchange scientists in Moscow are stark by American standards. The usual accommodations can only be described as primitive. Seasoned American travelers immediately become dismayed with the uncertainty of hot water, the lack of sink stoppers, the absence of bedsprings, and the interminable delays in the restaurants. Nevertheless, most take the cultural adjustment in stride as they begin to feel that they really are on an adventure. The following reflections of a few American exchange scientists indicate what life is like for American families in the USSR.

> Our reception and treatment were exemplary. All individuals we met did everything to make our stay pleasant and successful. . . . The people are great fun, are sensitive, and have no desire to blow up the Western world. . . . Everyone (except hustling cab drivers) was remarkably friendly, protective, and nice. Two young Soviet toughs were chewed out by a 55-year-old man for getting in front of us at a taxi stand and giving the wrong impression of Muscovites.

> It was either feast or famine, usually the latter. We ate American for breakfast with coffee, Tang, and Cheerios. The kids and I subsisted on sausage, bread, milk, and an occasional

small apple or green Cuban orange for lunch. . . . We knew we had sunk to our lowest ebb when the kids drank orange Fanta as the juice for breakfast. . . . The big treat at the university cafeteria was when they substituted tough steak for the mystery meat in the soup. It is fun to watch people picking up a steak and ripping into it with their teeth since it is impossible to cut the meat with two spoons.

Although the hotel was only 10 years old, the interior and the furniture seemed circa US 1930–1950. However, this gave a homey, well-worn feel to the place. . . . We were prepared to co-exist with an army of cockroaches, but we never saw one. . . . We saw no double beds. We used strong string to tie two single beds together, and we rolled up table cloths to fill the gap between the mattresses. . . . Russians think a dryer is the spin cycle of the washer. . . . The water was so dirty I sometimes wondered why I bothered to wash my clothes. I used only mineral water to brush my teeth.

Most chamber ladies sitting on each floor in the hotel were wonderful. After I gave them nail polish, stockings, and candy bars, their kindness became overwhelming. One of the ladies kept bringing us raspberry jam and honey.

Grocery shopping was surprisingly easy. Food was very limited, but there were self service push carts and two register lines. . . . I shopped before work, after work, and sometimes during lunch. I never knew when a prized item would appear on the shelves. I carried plastic bags and string bags so that I could stop at a moment's notice.

The transit system is outstanding. . . . The buses are so crowded that after everyone has squeezed aboard the doors can't close. . . . It is difficult to get off the bus at the right stop. I found remarkable incentive to learn such phrases as "!*%#! Let me off the bus."

Our children were definitely not young Pioneer material as they committed such lawless acts as dashing wildly down the hotel stairs, talking loudly, laughing, having temper tantrums at loud decibels, or wearing frail winter garments. . . . At restaurants and stores we frequently had to yield our daughter to eager strangers who would whisk her away to be shown off and cuddled. . . . The parks and playgrounds in Leningrad are ubiquitous and wonderful.

Our 13-month-old daughter jumped out of my backpack and landed squarely on her head. The taxi ride to the hospital almost killed us all. At the hospital the people were fantastic. . . . On another occasion, we saw five pediatricians. Each had a different set of diagnoses and medicines. . . . The hospital was somewhat self-service with mothers doing most things orderlies do in the United States. There were no charges for any medical services.

Either accept or decline an invitation; don't try to change the date since this is not the custom. Don't admire anything (like a picture or a book) too much, or it very likely will be given to you. If there are children in the home, bring small gifts for them.

The bath house is not to be missed. The steam bath itself, the ritual massage with birch branches, the beer and salted fish, and the feeling of participating in an ageless tradition all make it an experience to savor.

One of the most common daily events is to discover that a store, museum, or restaurant is closed for repair, inventory, or cleanup.

When I would answer a stranger that I was an American, the reaction was usually one of disbelief, followed by an excited handshake or animated conversation (80 percent) or extreme reservation and aloofness (20 percent). . . . Friends I made were real friends.[5]

When time for departure for home arrives, the visiting American scientists should be prepared for the hassles at the Moscow airport. Often upon arrival at the airport the scientists suddenly realize they have unspent rubles that cannot be taken out of the country. In principle, at the airport foreigners can change back to dollars those rubles that were obtained through exchange of foreign currency. However, this procedure to obtain dollars may take up to one hour since Soviet bank tellers seem to take their breaks at the busiest times of the day. Given the usual length of the lines to clear customs following any currency exchanges, few American travelers who have arrived at the airport only two hours before departure are willing to risk missing their planes by exchanging rubles. These unspent rubles usually are given to embarrassed Soviet hosts who promise to hold them until the visitors return.

The final step is the frantic search for the customs declaration prepared upon arrival several weeks or months earlier and the preparation of the departure declaration. The dollars declared upon arrival minus the dollars exchanged for rubles or spent in foreign currency shops during the stay in the USSR as evidenced by receipts should equal the dollars declared upon departure. Fortunately, these declarations are seldom scrutinized by officials at the airport. Few Americans are sufficiently meticulous in their personal financial accounting to pass a rigorous test of compliance with the formal procedures.

*   *   *

Soviet scientists arriving in the United States are usually more relaxed, since they know from their lifelong experiences with state stewardship in the USSR that someone will take care of them upon arrival. They are accustomed to long lines, to bureaucratic formalities, to baggage being lost, and to plans that never materialize. Finally, they know that the Soviet Embassy in Washington will take them in if necessary.

Soviet scientists usually arrive on Aeroflot flights that are booked solid with Soviet visitors to the United States. By the time the scientists have spent 15 hours en route to Washington or New York they have met other Soviets who are equally uncertain as to how they will be received in the United States. Any individual anxieties have become shared anxieties. Even those who have visited the United States before are given to excitement about the arrival and the days to follow, but they suppress outward manifestations of this excitement.

A bevy of Soviet officials and many American hosts regularly meet each Aeroflot flight. The pairing off of arriving Soviets with individual reception parties is always an interesting spectacle characterized by hugs, handshakes, and blank stares. The Soviets are usually unfazed by the customs formalities, which often involve long waits while American officials pay special attention to visitors from Communist countries.

Some American hosts for Soviet visitors are eager to present them with very detailed itineraries that have been cleared through many bureaucratic channels. Descriptions of the facilities they are to visit are frequently presented, misplaced, and even lost in the airport confusion.

Shopping in American stores is of the highest priority for many visiting Soviet scientists. K-Mart is among the favorite targets for Soviets eager to return home with presents for children, grandchildren, and occasionally spouses. Radio Shack has also become a favorite as electronic gadgets—which it is hoped include the appropriate adapters for the Soviet electrical system—become increasingly sought-after commodities.

Two years ago we arranged for a leading Soviet science official to visit Garfinkel's department store in the Washington suburbs, where he spotted a cashmere coat for $450. When he produced the money for the purchase, the sales clerk advised him to buy the coat at a nearby discount store. He thereupon paid $250 for the coat at the recommended store. We shared his bewilderment as to how American capitalism survives with such benevolent employees.

Naturally, shopping requires dollars. Therefore, it is not surprising that Soviet visitors eat many meals in their rooms, often relying on food brought from the USSR. Many Soviets save every possible dollar of their daily living allowances for gifts and souvenirs.

Soviet scientists coming to the United States dress in suits and ties. Seldom do they carry less formal attire. Their suits are remarkably resilient to all types of abuse, and they are not reluctant to wear the same suits on successive days to formal dinners and to picnics. The only awkwardness associated with their consistent apparel is the self-imposed embarrassment of the American hosts who are concerned that both Soviets and Americans are not dressed in the same style.

Recently I attended a formal dinner at the Department of State, where 200 Americans were dressed in tuxedos and evening gowns and the 25 Soviet guests were dressed in dark suits. Two days later I was hiking through the Pennsylvania countryside dressed in sports clothes together with the 25 Soviets wearing the same suits. They were totally relaxed on both occasions. Often the Soviets sense that their hosts are uncomfortable over the mismatches in apparel, and they take off their ties and jackets to appease their hosts.

Soviet scientists are well briefed on street crime in the United States, a type of crime that is uncommon in the USSR. They frequently are advised by colleagues in their institutes before departing from Moscow not to leave their hotel rooms after dark but to lock the doors and watch television. Elderly Soviet academicians often travel with younger colleagues, and frequently Soviet scientists travel in groups of two or more. This practice is due at least in part to concerns over physical abuse. There have been a few cases of muggings of Soviet scientists in major cities in the United States and Europe which have been used to support the continuous reporting in the Soviet press over the decadence of capitalism reflected in increasing crime rates. Also, from time to time visiting Soviet scientists leave valuables in unlocked hotel rooms or on park benches from where they quickly disappear.

For the first-time Soviet visitor to the United States, American reliance on the automobile makes a tremendous impact. When under the wing of a sympathetic American colleague, the Soviet scientist is mightily impressed with the advantages of private transportation. When on his own in the United States, the Soviet scientist is chagrined by the lack of well-developed public transportation systems and the high cost of taxis.

As we noted in an earlier chapter, Soviet scientists usually become instant celebrities when visiting small American towns. The hospitality is almost always overpowering, and the photographers are at every stop. Even hard-line local politicians begin to mellow in the euphoria surrounding the calls for peace and friendship. The Soviet visitors handle these situations very well, saying little and acting dignified. Sensing that they are being carefully scrutinized, they usually respond to questions in a straightforward and unassuming manner.

Perhaps the most candid reactions of Soviet scientists to their American experiences are obtained a few years after they return to the USSR. These delayed reactions are usually positive, with the most vivid recollections inevitably being linked to personal encounters with American scientific colleagues or with the man on the street. Also, I hear more comments about Disneyland in Moscow than in the United States.

During the visits to the United States, Soviets are simply overloaded with new adventures and quite uncertain as to the appropriate types of comments to make to newfound acquaintances. Why didn't the hotel clerk check my passport? Were those statements on annual salaries inflated? Do those children really know how to use the personal computers? They have previously read many accounts of the American way of life. Some of these accounts have undoubtedly been reinforced while others belied. They simply need time to sort out what they have experienced.

Time in the United States passes quickly. Now, chaos reigns at the Aeroflot counters during departures of Aeroflot for Moscow from the New York or Washington airports. This confusion must be a depressing experience, even for Soviet scientists who are returning home.

Soviet passengers have some of the world's largest and heaviest boxes and suitcases. I recently ordered two cars to transport several Soviets to Dulles Airport near Washington, only to find on arrival at their hotel that I had to pay $40 for an additional station wagon to take their five oversized boxes full of computers to the airport in addition to their luggage. In their hands they had leather bags taped together and stuffed with books, journals, and gifts which gave new meaning to the concept of carry-on luggage. At the airport, Soviet baggage clogs any attempt to establish orderly lines at the Aeroflot counters. Still, the Soviets assume that Aeroflot will not leave without them. They have never been wrong regardless of the delays in processing passengers.

*   *   *

In summary, both countries have very large pools of highly trained specialists working in almost every field of science. Many of their activities overlap; their approaches may be similar or different; and their conclusions may be complementary or contradictory. The scope of science is too vast and the resources of science too limited not to encourage the broadest sharing of ideas and experiences. While the practical problems of carrying out cooperative programs are formidable, the rewards can be substantial.[6]

CHAPTER 7

Trade, Cooperation, and the Techno-Bandits

No nation was ever ruined by trade.
Benjamin Franklin

Why do Soviet weapon designers copy Western concepts? The following Pentagon commentary wrestles with this question and sets the tone for Western efforts to protect high technologies which could feed the next generation of Soviet weapons systems.

In general, Soviet weapons have historically reflected a commitment to functional designs that can be easily manufactured in labor-intensive factories and readily maintained in the field with a minimum of technical skill. There has always been a struggle between Soviet design simplicity and technical complexity. Soviet weapon designers have not had to face the competitive pressures that drive Western designers to press the state of the art.

Building on a mature research sector and on lessons learned from past performances of weapons in battle, the Soviets are placing more of a premium on technically complex systems. Western system and equipment characteristics increasingly are used as yardsticks against which Soviet technical capabilities are judged. Every major civilian or military project is compared with the best Western technology before it is approved for development. Once in development, Soviet standards mandate the comparison of the quality and technical level of hardware, at different design stages, with foreign counterparts.

With their access to many details of Western weapons and dual-use equipment designs and concepts, Soviet designers are, in effect, competing with Western designers. That competition, supported and encouraged by the Soviet leadership, is probably pressuring the military research establishment to pay increasing attention to technically complex systems. Countervailing pressures for design simplicity are being applied by the manufacturing sector, which is less responsive in adapting to technological change. All of these forces indicate continuing Soviet programs to acquire Western military and dual-use hardware and technical data.[1]

*   *   *

Given the military technology competition between the two countries, many American skeptics doubt that scientific cooperation with the USSR makes sense. Some historians contend that Lenin prophesied that the Capitalist countries, in their zeal to exploit commercial markets, would sell the USSR rope with which the Soviets would in turn hang the Western Capitalists. Updating this prediction, American opponents of cooperation now ask, "Aren't we simply giving the Soviet Ministry of Defense a free ride by paying for research and development and then letting them have the benefits?"

One of the many critics of scientific exchanges is Richard Perle, a longtime congressional aide who subsequently became an Assistant Secretary of Defense during the Reagan administration. In his public comments since he left the government in 1987, Perle has prided himself on having prevented the liberals who had penetrated the Reagan administration from placing the nation in jeopardy because of their naive efforts to promote favorable relations with the USSR. His primary criticism of the government has been concentrated in the area of arms control. However, he also condemns those unenlightened former colleagues who "had no idea the Soviets were ripping off our technology so skillfully, so comprehensively, so effectively, right under our noses." He has urged the US Congress to listen to "the tooth fairy" rather than believe witnesses from the Reagan administration who argued that the US Government does indeed consider the risks as well as the benefits of cooperation with the USSR in science and technology. Perle has been very successful in sowing doubt about the value of scientific exchanges. He emphasizes the potential risks in exchanges and often ascribes technology losses to exchanges that in fact occurred through illegal diversions of trade in sensitive items.

Perle's message has been very simple: We must remain technologically superior to the USSR in militarily critical areas to deter their expansionist ambitions. Others who share his views then argue that many fields of science and technology will eventually have military applications; therefore, militarily critical technologies should be very broadly defined to include even those areas of basic science which within 10 to 20 years could possibly spawn military applications. The argument concludes that the Soviets should be denied access to such research now to ensure protection of our technology later.

The Soviets will spare no effort to obtain our technology since they have little capability to develop their own; thus, science and technology exchanges are an important avenue for Soviet efforts to obtain important information, according to Perle. Administrators of American exchange programs are repeatedly duped, he contends; and he notes that they simply don't understand that Soviet participants in exchanges are really intelligence agents in disguise, that within the USSR there are shadow laboratories that American exchange participants never see, and that the Soviets have little to offer anyway.

This position attracts considerable support in the US Congress and throughout the country. Even "liberals" do not want to appear soft on communism, and technology transfer is an easy target for Soviet bashing. Indeed, Perle's position reflects stereotypes of the past; it contains sufficient truth and is reinforced from time to time by incidents of Soviet efforts to steal Western technology that few politicians have the inclination to explore the issues beyond his sweeping conclusions.[2]

*   *   *

A plethora of US Government documents and many public pronouncements are designed to support the government's denial of Soviet access to many aspects of American science and technology. The Department of Defense has issued unclassified documents and then translated them into several languages to warn about the nefarious activities of the Soviet intelligence services in acquiring Western technologies for their military forces. The Pentagon has focused on American universities as well as industrial facilities as points of leakage of important information, and particularly those universities which have been awarded contracts and grants to work on SDI projects.

The FBI has enthusiastically supported the calls by the Pentagon for greater strictures on visiting Soviet scientists. For example, in 1988 the FBI released a brief report warning how the Soviets were exploiting American libraries. Librarians were asked to keep lists of visitors with foreign names and foreign accents which could be used to identify Soviet agents. While the FBI elicited the cooperation of some librarians, many strongly objected to this tactic and voiced their complaints during television interviews. The FBI accusations were very general and cited only one specific case of a Soviet diplomat systematically microfilming unclassified library documents and stealing some of the documents.[3]

More convincing evidence of Soviet intentions does exist. In the early 1980s, French intelligence sources obtained a classified Soviet document which describes the efforts of Soviet intelligence agencies to exploit Western technologies in the field of aeronautics. This authentic documentation leaves little doubt that the Soviet "collection" program, at least in the field of aeronautics, was extensive at that time. The effort relied on legal and illegal means, open and clandestine approaches, and Soviet and foreign agents to obtain information. The program targeted industries and universities, institutions and individuals, and classified and unclassified projects in the United States and Western Europe.[4]

According to Pentagon analyses, exploitation of Western technology, and particularly American technology obtained through illegal trade diversions, has saved the countries of the Warsaw Pact billions of dollars, reduced the time required for developing their weapons systems, enhanced their defense industrial productivity, and allowed them to respond more promptly to Western weapons systems and military tactics. Such tactics have forced the United States to make additional financial expenditures in a response to enhanced Soviet capabilities. These same analyses underscore that the losses would have been far larger had not the American defense community been sensitive to the Soviet activities and taken appropriate countermeasures.[5]

An American intelligence conclusion which has been widely quoted by the Pentagon is:

The acquisition of Western technology and finished products is a very calculated procedure done with a great deal of selectivity. Even so, Soviet economists are amazed that the West does not recognize this. They consider this acquisition program one of the USSR's greatest achievements since it allows the solving of complicated problems with minimal costs.[6]

These types of conclusions are buttressed by CIA commentaries such as the following:

In the development of microelectronics, the Soviets would have been incapable of achieving their present technical level without the acquisition of Western technology. . . . Their advance comes as a result of over 10 years of successful acquisition—through illegal, including clandestine, means— of hundreds of pieces of Western microelectronic equipment worth hundreds of millions of dollars to equip their military-related manufacturing facilities. These acquisitions have permitted the Soviets to systematically build a modern microelectronics industry which will be the critical basis for enhancing the sophistication of future military systems for decades.[7]

As to further evidence, Soviet SS-13 missiles look strikingly familiar to the designers of our Minuteman systems. Their Advanced Early Warning aircraft resemble our AWACS designs. Their SA-7 heat-seeking, shoulder-fired antiaircraft missile contains features of the American Redeye missile. On the civilian side, their shuttle has a NASA look, and the Il-76 aircraft incorporates many features found in the Boeing 747. The Pentagon's favorite example of Soviet technology obtained through covert collection is the Soviet's look-down shoot-down radar system, which is patterned after the fire control radar of American F-18 jet fighters. Also of particular concern has been the improved accuracy of Soviet land-based missiles, believed to have benefited from acquisition of Western gyroscopes, accelerometers, and other guidance components.[8]

In recent years the American press has widely reported Soviet attempts to obtain specific technical information and hardware of military relevance from the United States. Many popular media stories in California describe the Soviet techno-bandits riding unconstrained along the highways and byways of Silicon Valley in search of the vulnerable Americans who have expert knowledge of megachips of the future. These stories are usually far less worrisome, at least to Californians, than the stories of Japanese efforts to obtain foreknowledge of the next generation of American products which could be readily manufactured under Japanese labels. Economic competition affects the people; military confrontation affects the politicians, conclude the residents of Silicon Valley.[9]

Espionage and other undercover approaches for acquiring technology have become quite sophisticated. The traditional revelation by the press of a simple plot for a Soviet diplomat to receive an unmarked envelope at a letter drop from a collaborating American engineer has given way to intriguing literary essays on far more complicated technology transfer networks. They involve multinational companies, transhipment of sensitive hardware in and out of distant free ports, and eventual reproduction of engineering designs in the USSR.

The discovery in early 1988 by the US Government of the sale to the USSR by Toshiba and other Japanese manufacturers of advanced technologies for production of quieter Soviet submarines set off a firestorm in Washington. American politicians immediately threatened trade sanctions and other retaliatory measures against Japan. The incident compounded the growing bitterness toward overall Japanese trade practices, which were considered to be discriminatory toward the United States. However, the military significance of this transaction became confused as conflicting statements were issued by the Pentagon. On the one hand, senior Pentagon officials stated publicly that the technology had previously been available to the Soviets. Yet, on the other hand, other Pentagon officials were bemoaning the billions of dollars that would be required to counter the advantage gained by the Soviets through the Toshiba purchase.[10]

In short, there can be no denying that the Soviet Union has targeted Western technologies as valuable ingredients for its industrial and military modernization. Indeed, a central thrust of Gorbachev's perestroika is outreach to the West for science and technology, and the Soviet intelligence authorities simply applaud and say, "We've anticipated your brilliant strategy. Just turn us loose."

*   *   *

How serious is the technology loss? When the Japanese steal our technology, the consequences soon become apparent. Jobs are lost, and our balance of trade suffers. The international economic competition is measured in dollars and cents and directly affects the well-being of our entire country.

When the Soviets steal our technology, the consequences are far less obvious. Americans face no economic threat from the USSR, at least not for the next several decades. Enhanced Soviet military power through the improved performance characteristics of their armed forces is very difficult to place on a scale of significance, particularly as arms control and unilateral restraint become more important in the strategic relationship between the two countries. However, American military planners will continue to be greatly concerned over Soviet submarines that run a little quieter, missiles that have greater accuracy, tanks which have greater firepower, and radar sets that do not give false identifications.

Despite the euphoria surrounding the four Reagan-Gorbachev summits, the United States and the USSR will be political adversaries for some years into the future, with military

power used to support political positions. Within this framework of political reality, Soviet efforts to ensure that their military forces have the latest technologies take on considerable
meaning for the American public. A large fraction of our tax-
payer dollars supports a defense budget based on the need to
keep ahead of the Soviets technologically. Quieter Soviet submarines mean greater expenditures to improve our antisub-
marine warfare capability; improved Soviet radar capabilities
trigger new programs to enhance the stealth bomber; and better
accuracy of Soviet missiles translates into new techniques for
hardening our communication centers.

The significance of such military measures and countermeasures—given the large size and extensive capabilities of current
arsenals in the two countries—is largely a matter of judgment of
the political implications of improved military capabilities. In the
past this significance has been argued primarily in military terms
through complicated military scenarios that often divert attention from the importance of political will as the basis for military
aggression or restraint.

As we have seen, US determination to stay ahead of the
Soviets technologically, regardless of the cost, is clearly reflected
in SDI. Given the political commitment of 1983 to development
of an effective defensive shield, the US Government became
very sensitive about the transfer of any type of American technology or of any information about the details of SDI that might
contribute to Soviet capabilities to overwhelm such a system.
This position was an ironic about-face from Ronald Reagan's
initial offer in 1983 to share SDI technology with the USSR.

Meanwhile, the Soviets will clearly develop the most effective and most sophisticated weapons systems that they can,
with or without assistance from Western technology. Is it better
for the Soviets to be technologically dependent on the West than
for them to develop stronger indigenous capabilities? If they are

dependent on the West for designs and hardware, won't they always lag behind, and won't the Western powers have a good basis for judging the directions of their weapons efforts? If, however, they are developing their own capabilities, couldn't they surprise the West with unexpected approaches?

Strong American critics of technology sharing call this type of thinking absurd, since they do not believe that the Soviets would independently achieve breakthroughs unknown to the West. They dismiss Soviet technological successes in space research and in fusion research as isolated events during an earlier period of history. Now our high-technology industry will keep us well ahead of the Soviets; and our strengthened intelligence sources will prevent surprises, they argue.

How important are the cost savings that accrue to the USSR through the use of Western technologies? These technologies include highly specialized military technologies, dual-use technologies that could be used in either the military or civilian sectors, and technologies that are used only in the civilian sector. By not having to reinvent products in areas in which the United States has a decided lead, the Soviets should be able to cut costs in both the military and civilian sectors by exploiting American research and development.

Historically, the Soviets have never skimped on essential military expenditures. During times of economic stringency, they have achieved cost savings through limitations in civilian areas, particularly in the consumer goods sector. While a relationship obviously exists between the overall health of the Soviet economy and the level of military expenditures, cost savings from acquisition of technologies from abroad will not be the decisive factor in determining the size of military budgets in the USSR.

The question should not simply be whether Western technologies contribute to cost savings for the Soviet military establishment. The Soviets have always retained a force level considered essential regardless of the financial pain. Thus, more broadly, should blocking Gorbachev's achievement of a more prosperous Soviet economy be a significant factor in determining whether American technology should be denied to the Soviets, recognizing that the larger the economic pie, the more resources there will be for all sectors including the military sector? This issue antedates Gorbachev by many years with the previous phraseology being, "Do we prefer a fat Russian bear to a hungry Russian bear?" The question takes on particular relevance now since Gorbachev's survival, together with the survival of many political reforms in the USSR and recent Soviet arms control initiatives, depends heavily on whether he succeeds in the economic arena.

During the next decade and probably well into the future, international thievery of technology, and particularly technology which originated in the United States, will be of growing concern among many segments of the American public. The Soviets have many sophisticated approaches to obtaining technologies developed by their adversaries. Nonetheless, from time to time they will undoubtedly continue to be caught "red-handed" in their covert efforts. Soviet attempts to obtain embargoed items in violation of American laws will be a major impediment to developing mutual trust between the two countries, with serious ramifications for improving relations on many fronts.

In short, the military and economic dimensions of the Soviet exploitation of our technology advances are important, but the political dimension is much more significant.

*   *   *

The Soviets are not alone in seeking technologies developed abroad. Organizations in every country, including the United

States, are constantly looking for technologies developed around the world which can be used to save time and money. Nevertheless, contrasts abound in the American and Soviet approaches to the acquisition of foreign technology.

The American approach relies primarily on private firms. Thousands of American companies have tentacles into many countries to assess technological as well as marketing developments in their fields of interest. These international linkages expand each year and provide invaluable information into promising trends worldwide. Most major American firms have a good grasp on the principal foreign activities of potential commercial interest.

For decades, the US Government has been trying to help American companies capitalize on foreign technological developments. For example, the US Government, through the Department of Commerce, provides American companies just entering the international arena with leads on potential foreign commercial partners, leads often identified with the help of American embassies abroad.

Few American firms are motivated by the likelihood of finding useful technologies in the USSR or Eastern Europe. Rather, the American companies interested in the region seek to use American technologies as a bargaining tool to break into a large undeveloped market or to gain access to raw materials. However, in the course of their international dealings, American companies become keenly aware of many technological developments in these countries—even in countries as tightly controlled as East Germany and the USSR. Some of the information they acquire may be of considerable interest to military authorities even though such findings are quite incidental to their search for commercial opportunities.

While private companies possess most of our nation's technological wherewithal, the US Government plays a central role in determining requirements for military hardware. These performance requirements are developed within the context of intelligence estimates of the Soviet "military threat," and the US intelligence agencies expend a great effort in acquiring information concerning that threat, including assessments of current and future Soviet technological capabilities. Technical specifications for military hardware are then determined by the Pentagon. Finally, private companies play a critical role in designing the approaches to meet the requirements, using their technological capabilities that could have originated anywhere in the world.

American intelligence agencies describe the centralized Soviet program to acquire foreign technology as massive. It involves large numbers of people and large sums of money dedicated exclusively to the task of "technology collection." The combined efforts of American companies to draw on foreign technological developments cannot be measured but is also massive. Since American companies are an integral part of the international technological community, they do not need to exert an extra effort to pursue the types of technological information for which the Soviets expend an enormous effort.

As a result of the contrasting approaches, the Soviet technology collection effort has high visibility in the West and particularly in the United States. The Soviet program is centralized, is orchestrated by the government, and relies to a great extent on the Soviet intelligence agencies. These agencies are also charged with obtaining information on the military threat to the USSR. The US approach generally separates these two functions, as outlined above: our intelligence agencies assess the threat and American private companies stay abreast of worldwide technologies to be in a position to respond to the threat. Also, one of the prime targets of the Soviet technology collection effort is the United States, whereas, as we have noted, American companies have little interest in Soviet technology and concentrate their overseas searches for technology in Western Europe and Japan.

The Soviets rely to a great extent on covert acquisition techniques, while the American companies engage in what we define as acceptable commercial practices. For example, Western trade fairs are considered usual business undertakings, while Soviet trade fairs have a very checkered record in mixing commercial and covert intelligence activities. In December 1988, I attended a trade fair for advanced scientific laboratory equipment housed in three large pavilions in Moscow. The extensive European and Japanese exhibits were quite sophisticated, particularly their computer software. The American exhibitions were very modest. The Soviet research institutes presented an impressive variety of their technical achievements. This fair could be viewed as a Soviet effort to entice unwitting Western business executives to bring to Moscow high-tech items which might otherwise be subject to licensing or simply as a Soviet effort, Western style, to sell and buy products. It may have been a little bit of both.

The Soviet technology acquisition program can also be viewed as an effort to compensate for the isolation of Soviet specialists from Western colleagues and from the normal international flow of unclassified information. In particular, many Soviet travelers abroad want to go to the copy machine with almost every scientific document they see, lest they never see it again. Of course this practice makes a very bad impression on Western hosts regarding Soviet motives.

*   *   *

How does the Soviet technology collection process work? A special commission within the Soviet Government plans the exploitation of international technologies, including American technologies, with various ministries feeding requirements to this commission. These requirements are collated and sorted and then assigned to appropriate Soviet organizations for collection. This reliance on intermediary organizations to service the researchers and engineers who are not directly tied into the international flows of technical information is bound to be highly inefficient. The sheer magnitude of the effort prevents collection of information carefully screened to satisfy the needs of many individual scientists whether they be working in the military or civilian sector; an exception may be a requirement simply to acquire hardware or blueprints.[11]

Recent publications of our Department of Defense highlight the role of the Academy of Sciences of the USSR itself as one of the technology collection organizations of the Soviet Government.[12] Indeed, these reports describe the Soviet Academy as an intelligence agency, and Mr. Perle strongly criticizes the enthusiasm of American scientists interested in cooperative programs with the Academy. The Academy does play a prime role in Gorbachev's outreach to the West. Though Soviet intelligence agencies such as the KGB undoubtedly influence some of the international activities of the Academy, this is a far cry from describing the Academy as an intelligence agency. Relatively few of the tens of thousands of scientists of the academy system who meet with Western scientists come into the kind of contact necessary to allow them an opportunity to participate in gathering intelligence information. While some scientists who participate in international activities may have assignments from the intelligence agencies to collect material that is accessible, they also usually have important scientific reasons for their involvement in international scientific contacts. When it comes down to the realities, the activities of the Soviet Academy are overwhelmingly scientific and not as sinister as they have been portrayed.

Many departments and technical agencies of the US Government also have international programs which provide useful information to the American intelligence community, and indeed our government takes pride that the activities of various government agencies are well coordinated. Sometimes scientists from our technical agencies may be alerted to specific types of information of interest to the intelligence community, including information concerning the cost of electricity, results of planetary missions, pollution problems, and many other activities far removed from classified military programs. Surely the Department of Energy, NASA, and the Department of the Interior are not intelligence agencies. While Soviet intelligence efforts may not be equally benign, we should not be too quick to condemn every Soviet scientific outreach as inspired by intelligence agencies.

The type of rhetoric used by Mr. Perle and others to paint a foreboding picture of Soviet technology collection efforts is surely effective in deterring more cooperative efforts. Unfortunately, it too often ignores the legitimate international interests of Soviet scientists and inhibits progress toward mutually beneficial exchanges.

*   *   *

Despite the foregoing concerns, above-board scientific exchange programs between the two countries have been effectively carried out for more than 30 years. Initially, the USSR had used these programs to help identify and enlist Western firms that could provide the USSR with needed technologies. During Soviet-American exchange negotiations throughout the 1960s, Soviet officials were quite candid in expressing to me and to other American officials their interests in acquiring Western computer technologies, for example, through formally structured exchange programs. Specific exchanges in this and other high-technology fields were negotiated as part of the overall exchange packages. Some of the exchanges eventually led to Soviet purchases of equipment and licenses.

In agreeing to these early exchanges in high-technology fields, the US Government considered that brief exposures by Soviet specialists to American achievements which were widely reported in American technical journals would not jeopardize our national security. For our part, we wanted to visit their facilities since we had almost no authoritative information about the state of Soviet developments. I remember accompanying an American delegation to a computer factory in Minsk in 1964, where we saw hundreds of Soviet women squinting through magnifying glasses as they assembled integrated circuits, a practice abandoned in the United States a few years earlier. Our specialists from IBM and Bell Laboratories who participated in the visit simply could not believe they were in the same country that had launched Sputnik seven years earlier. This visit was important in helping to put into perspective the "missile gap" theory being propounded by the Soviets and the Washington agencies. To catch up with the Soviets, US officials asserted, they needed larger budgets.

During the 1960s, the Soviets probably had some success in identifying important technologies during exchange visits, acquiring them commercially, and then incorporating them into military systems. The Soviets frequently produce military systems on a tightly controlled customized basis in contrast to mass production of civilian items, thereby easing adaptations from abroad. However, the military systems of 20 years ago were far less complex than today, and consequently integration of Western and Soviet technologies was less complicated in the past.

Exchange programs gradually lost most of their significance as a channel for transfer of technologies needed in the Soviet military or industrial efforts: easier and more direct routes to

American technology developed. East-West commercial contacts now involve hundreds of organizations and firms in both countries despite the small amount of trade. Joint Soviet-American staffs are in place in Moscow and Washington with the sole purpose of helping American firms and Soviet organizations expand their commercial relationships. These commercial relationships involve substantial cooperation in science as well as technology, and particularly in science that has applications within the next decade. Similarly, the Soviet network for monitoring military research and development activities in the West is well established—operating through diplomatic and clandestine channels. The periodic arrests of Soviet and American participants in these activities provide ample evidence of the aggressive nature of the network. Indeed, when it is revealed that American military personnel with access to highly classified activities in very sensitive fields have been passing high-technology secrets to Moscow for a period of years, some of the attention given to the possible loss of information through scientific cooperation becomes more difficult to understand.

Given US concerns over the outflow of technology to the USSR, the United States currently emphasizes science rather than technology in exchange programs and, for example, tries to avoid any relations involving large computers. The Soviet appetite for exposure to American technology remains boundless, and Soviet officials continue to state openly their interest in using exchanges to gain access to US technology. Though they are not happy with the US emphasis on science, they have reconciled themselves to the unwillingness of the United States to cooperate in high-technology fields; and they have simply increased their reliance on other channels for collecting technological information.

An incisive review of US-USSR exchange programs carried out by an independent panel of the National Academy of Sciences in 1982 showed that even when technology was an essential component of exchanges, only a very small portion of the flow of militarily relevant technology to the USSR could be traced to exchange programs.[13] These conclusions remain valid today as less and less of the contact between Soviet and American specialists is channeled through exchange programs. As noted previously, there simply are many other more efficient methods for the Soviets to obtain technical information.

If exchange programs are not primary channels for the flow of militarily relevant technologies, why do the Soviet intelligence agencies, including the military agency, the GRU, continue to assign Soviet exchangees the task of searching out information for them? Inertia from collection efforts of the past? Any speck of information is helpful? Looking for potential American intelligence collaborators? To identify the existence of technologies which should be future targets for intelligence collection through other means? Probably all of these reasons. The intelligence mind-set conveyed to agents abroad is, "If it is free, grab whatever you can; and we'll sort it out when we receive it."

While the US Government is becoming increasingly preoccupied with the outflow of sensitive information through scientific exchange programs, despite the questionable importance of these channels, the governments of Western Europe and Japan apparently are less concerned. Frequently, the US Government does not approve exchange visits to the United States for Soviet scientists, unaware that very similar exchanges are going forward between research institutes in the USSR and laboratories in other Western countries which have the same level of sophistication as the American laboratories. Indeed, Soviet access to comparable scientific methodologies and information in other countries is seldom considered in reviews in Washington of proposals for Soviet-American scientific exchanges.

*   *   *

Let us now turn to how the United States has responded to Soviet techno-banditry by restricting Soviet access to American scientific and technological accomplishments. For more than three decades the US Government and the governments of our European allies have joined together to limit the flow of engineering products and designs of military significance from West to East. Maintaining our technological edge will let us "fight smart," notes the NATO leadership.

After World War II the United States controlled most modern military technology. Americans invented submarine detection systems, over-the-horizon radar, solid rocket propellants, and many other technologies. Our diplomats usually dominated discussions among Western governments concerning the specific technological designs, components, and test equipment which should be embargoed. The US position in safeguarding our technological inventions has consistently been more protective than the positions of our allies. The United States, and not they, shouldered the technological responsibility for the defense of the Western alliance. Americans paid the costs of research, and our government naturally has been possessive of the results of these investments. It resists giving to Soviet military competitors new ideas which could require additional investments on our side to counter their adaptations of our technologies.

However, in recent years scientific and technological advances in other countries have eroded US dominance as the unique reservoir of many advanced Western military technologies. France, the United Kingdom, West Germany, and other NATO countries have made substantial technological contributions of military interest. In some areas they have simply caught up to where we were some years ago, but even in these areas their achievements have current military value. British sailors certainly learned about the effectiveness of French missiles of the 1970s mounted on Argentine aircraft during the recent Falklands war. In some areas such as the design and testing of nuclear weapons, our allies are working near the cutting edge of modern military science. Thus, the US Government now takes their views on the specific details toward protecting Western military technologies from exploitation by the USSR more seriously.

Western attitudes toward the flow of technology to the USSR in nonmilitary areas have always been quite divergent, with the West Europeans consistently being more interested in East-West trade involving advanced technologies than the United States. The attitudes have changed as the Cold War heated and cooled and as diplomatic, economic, and espionage incidents marred the bilateral relations of individual countries with the USSR.

These differing views erupted when the West Europeans decided to assist the Soviets construct a gas pipeline across the western part of the USSR. The pipeline will eventually reach Western Europe. US concerns over pipelines across the USSR which could feed Eastern and Western Europe date to the early 1960s. Indeed, in 1965 I prepared a diplomatic report on the state of pipeline construction across the western USSR using information I had stumbled upon during casual conversation with drunken Soviet army officers on a train from Moscow to Lvov on the Polish border. The Soviet soldiers had become intrigued by an American who seemed to speak Russian with a Latvian accent and who was conversant about the achievements of the Soviet national basketball team.

The latest clash of Western policies marred the 1985 economic summit of Western heads of state in France. Despite former President Reagan's impassioned pleas, the Europeans ignored the American objections that technologies relevant to the construction and operation of the pipeline should be embargoed by the West, and they concluded a variety of commercial arrangements with the USSR. Within about one year, the US Government changed its opposition; but it was too late for American commercial interests to share in many of the economic benefits of trade arrangements.

In recent years the issues surrounding the control of Western technologies have become increasingly complicated and volatile as the lines between military and nonmilitary high-technology products rapidly disappear. Diversion of civilian technologies to military purposes is often frightfully easy. Also complicating control has been the diffusion to all continents of many technologies of interest. Ships from Soviet bloc countries taking on large cargoes of electronic equipment in the harbors of Singapore and Macao, and Hungarian trucks with loads of computers traveling from Vienna to Budapest have become common sights.

Controls on technology originating in the United States have sparked lively debates on Capitol Hill, where they have been incorporated into several export control laws administered by the Departments of State, Commerce, Energy, and Defense and by NASA. Each year as military systems become more complex, the laws become more complicated in their interpretation, their administration, and their enforcement.[14]

The divisive issue, as we previously noted, is not the philosophical question of whether we should deny Soviet access to our military technology. We should. Rather the issue centers on the criteria used in identifying those developments which should be protected and the details of how they are to be protected. While the Department of Defense argues for embargoes on the maximum number of items, the Department of Commerce urges reducing the embargo list lest American trade opportunities in the civilian sector be jeopardized. The White

House and the Department of State are swayed more easily by twists and turns in our political relations with the Soviets: we should punish them by embargoes in bad times and reward them by easing embargoes in good times. Also, the State Department searches for compromises among Western allies which balance risks and benefits while also taking into account political realities and longer-term US interests.

During recent years the US list of embargoed items steadily grew. The need to protect technologies related to nuclear weapons, military armaments, and electronic detection devices has always been clear. But now the control list covers items in many commodity groups including metal-working machinery, chemical and petroleum equipment, electrical and power generating equipment, general industrial equipment, transportation equipment, electronics and precision instruments, metals and minerals, chemicals and petroleum products, and rubber and rubber products.

The enlarged staffs of the Pentagon, assigned the task of identifying potentially worrisome technologies, began to take control of the technology transfer discussions in Washington and of discussions with European allies during the 1980s. American commercial interests reacted aggressively and began documenting losses of trade opportunities to competitors in Western Europe and Japan in the billions of dollars each year as other countries relied on much shorter lists of embargoed items. Meanwhile, our diplomats tried their best to build a consensus on individual items through the Coordinating Committee on Export Control (COCOM) in Brussels, where the Western allies and Japan develop common positions on embargoes.

Restrictions on the export of embargoed American products to our allies became a particularly contentious issue in the 1980s since these restrictions cut sharply into the sale of American goods in Western Europe. Much of the problem was attributed to delays in processing export applications. Also, when an American ships certain embargoed items to a customer in Western Europe, the customer may be required to provide documentation that the item will not be reexported to a Communist country or to another country where further reexport could lead to Moscow, Prague, or other East European cities. This American requirement for reexport certification is not matched by any other country, and it has jeopardized the quick response capabilities of American firms in competing for trade opportunities.

An approach to export control advocated by several American study groups and increasingly supported by the US Congress calls for a greatly reduced list of embargoed items but more aggressive efforts to seek international consistency in ensuring the effectiveness of the embargoes. The embargoes should be directed primarily to items which could significantly enhance Soviet capabilities to *manufacture* military hardware, with much less concern devoted to controlling individual end products which have been manufactured in the West. New Soviet production capabilities in highly specialized areas pose a much greater military threat than Soviet access to complicated end products which require elaborate reverse engineering efforts to figure out how they were manufactured and then to develop the manufacturing capability.[15]

A second approach which has received less attention would call for protecting the "lead times" necessary to develop new military systems. No system can be protected forever from acquisition by an adversary, and unclassified technologies cannot be protected for very long. Rather than indefinitely protecting specific technologies, which is often the current approach, the United States might be better served by ensuring that lead times of perhaps 10 years are maintained over the USSR in selected high-technology fields. The strategy would be to concentrate on

controlling technologies developed during the past decade coupled with continuing research efforts on industrial improvements which will enable us to maintain our lead. Of course, a
few older items might be of continuing concern, but the bulk of
the controlled items would be of recent vintage. Though older
systems can frequently pack a mean punch, once they have
been deployed for some time, effective control becomes increasingly difficult.

*   *   *

Restrictions on exports to the Soviet bloc of industrial design information, manufacturing equipment, manufactured
goods, and related technical know-how have historically been
the focus of efforts to control Western technology. Now, as reflected in the comments of Richard Perle, limitations on Soviet
access to scientific information as well as to technology developed in the West are of more intense interest within American
military circles than ever before.

The dilemma is vexing. As long as the Soviets are our military adversaries, shouldn't they be denied access to scientific
information which could eventually benefit their military effort?
But the freedom to publish and exchange information is a tradition as old as science itself and is one of the greatest strengths of
our scientific enterprise.

This dilemma is not easily resolved. In the age of high technology, many scientific developments precede applications by
only a few months or a few years. Most scientific information
could conceivably—over the long term—be of interest to military planners. Our military forces employ or enter into contracts
with specialists from every branch of science, and the Soviet
armed forces undoubtedly also draw on all branches of science.
On the other hand, the practicality of freely circulating scientific
information developed in the United States to scientists in some

countries but withholding it from scientists in other countries is questionable at best.

The financial argument again lurks in the background. Some government officials have strong convictions that the United States simply should not help the Soviets economically as long as they pose a military threat. The United States should not help reduce the cost of their scientific research by letting them capitalize on American research results or by allowing them to identify blind alleys that have already been explored by American scientists, they contend. They add that applications of scientific research usually have economic payoffs in addition to any military payoff. While the technology transfer discussions should be directed to military concerns, the economic dimension must not be ignored.

Almost all American scientists accept the concept of *classified* information. Those scientists in the United States and in other Western countries working on secret projects recognize a need to limit distribution of information which they develop. Restrictions on *unclassified* scientific information trigger intense debates, since scientists depend on information developed by colleagues around the world and resist governmental intrusion that limits international contacts.

In the area of export control, the regulations are gradually widening to include unclassified scientific information in certain areas such as artificial intelligence and biotechnology as well as industrial information which has always been controlled. Limitations on exports of computer software are expanding, and they increasingly impact on cooperation as computer programs become standard tools in many scientific disciplines. But software packages are so numerous and so versatile that distinguishing those which could provide the Soviet military establishment with significant new capabilities is often impossible.

The Department of Defense has been placing more stringent limitations on unclassified information resulting from their contracts and grants. This trend troubles some American researchers since the Department of Defense now controls a large portion of federal support of basic research. For many unclassified programs sponsored by the Department of Defense, such as the large unclassified SDI basic research program carried out at about 60 American universities, the American researchers must seek approval prior to releasing information developed through their projects. As would be expected, such approval has been withheld in some cases. While researchers usually prefer funding with strings to no funding, many strongly believe that limiting the flow of unclassified information is unnecessary and reduces opportunities for advancing science.

Restrictions on the activities of American scientists with security clearances or on scientists working on government-funded projects have also created anxieties. The US Government sometimes prohibits their travel to Eastern Europe or restricts their participation in unclassified meetings in the United States which involve contact with scientists from Soviet bloc countries. The frequency of such limitations is of course important. But perhaps of even greater impact is the chilling effect of the approval procedures. Most scientists are not looking for hassles, particularly with the organizations which provide their funds. Therefore, some prefer not to try to participate in programs that might require special approvals or could involve questions concerning their judgment or their loyalty to the United States.

Since every Soviet scientist visiting the United States must have an American entry visa, the US Government has a powerful tool for controlling scientific contacts in this country. The archives of the Department of State bulge with hotly debated decisions as to whether Soviet visitors to the United States will

be exposed to American scientific achievements which can significantly improve Soviet military capabilities.

In general, the process of the US Government for reviewing
visa applications for scientific visitors from the USSR is reasonable. Upon receipt of a visa application from a Soviet scientist,
the Department of State seeks an advisory opinion from the
American intelligence community as to whether the visit will
result in unacceptable losses of scientific information which
could help the Soviet military effort. With this opinion in hand,
the department then decides whether a visa should be issued
and whether limitations on activities in the United States are
appropriate, also taking into account any political aspects of the
proposed visit.

However, in the development of the advisory opinion, the
technical resources of the Department of Defense often overwhelm the other intelligence agencies, and that department understandably takes very conservative positions. Even if Soviet
visitors will not have access to sensitive information, they will
be able to identify sensitive programs for subsequent Soviet
penetration by clandestine means, the department may argue.
On occasion other agencies might add: visitors will assess the
personal vulnerabilities of their American hosts, which can be
exploited when the American hosts travel abroad; they will record the serial numbers of scientific equipment which can then
be obtained illegally, or they will charm their hosts into talking
too much and revealing technical data in violation of export
control laws.

Will visa decisions to deny Soviet participants in scientific
exchange programs access to unclassified activities really make a
difference in protecting the national security of the United
States? As previously discussed, the Soviets have many other
channels for obtaining sensitive information. However, given
the political volatility in Washington of the issue of protecting

American technology, the US Government vigorously pursues visa control as a device for limiting Soviet access to American science. In 1983 the US Government abandoned visa control over visiting Chinese scientists and engineers as a means of protecting American technology after the Department of State reversed 65 negative intelligence advisories on political grounds. But, the vision of the Soviet military threat is not as easily dismissed.

Thus, stringent visa control for Soviet scientists visiting the United States will remain. American critics of the value of scientific exchanges will continue to emphasize that many Soviet visits to American universities and other research centers are fraught with danger for our national security. At the same time, the United States must strive to preserve the benefits of genuine cooperation, and a balance must be reached between the promise of international science and the perceptions of national security.

Therefore, criteria for determining the appropriateness of restrictions on scientific exchanges seem essential. Such criteria were developed in 1982 by the panel established by the National Academy of Sciences at the request of several US Government agencies to review the national security aspects of scientific communication, which was previously noted. The panel recommended that restrictions are warranted only if the scientific field in question meets all four of the following criteria:

- The technology is developing rapidly, and the time from basic science to application is short.
- The technology has identifiable, direct military applications.
- Transfer of the technology would give the USSR a significant near-term military benefit.
- The United States is the only source of information about the technology, or other nations that could also be the source have control systems as secure as ours.[16]

Some Soviet proposals for scientific cooperation are probably designed to test the limits of US technology transfer policy, and they fall within the area of concern to the panel. On the other hand, the US Government has not followed the very sound advice of the panel and has rejected many proposals of Soviet and American institutions for cooperation in fields clearly outside the area of concern to the panel. While there are difficulties in relating specific fields of interest to scientists in both countries to the very general criteria, greater efforts are needed to use the criteria noted above in reviewing the technology transfer aspects of proposed cooperative programs.

*   *   *

Obtaining research results, blueprints, or equipment incorporating foreign technologies is but the first step in a long and complicated engineering task of effectively using the foreign achievements. The Japanese are usually looked upon as the masters of this task. But in many fields the ingenuity of American entrepreneurs in mating foreign technology with indigenous achievements is unsurpassed, and many products on the American market reflect sophisticated integrations of American and foreign technologies. Just examine the products on sale at Radio Shack.

For the Soviets, however, the task of blending Western and indigenous technologies is often insurmountable. In some cases the Soviets have simply purchased turnkey plants and relied totally on the Western approaches rather than attempting to incorporate the foreign technology into their manufacturing. Given the size of the task of upgrading their industrial base, the Soviets must now draw primarily on their own capabilities. Foreign achievements are not a substitute but can contribute to improving and refining the achievements of their own engineers.

For the Soviets, the problems of technology transfer are manyfold. First, as mentioned before, Soviet engineers are usually divorced from international developments and frequently can only guess where to look abroad for many details of interest. They then must rely on Soviet intermediary organizations (e.g., Soviet intelligence agencies or foreign trade organizations) with limited technical perspectives to go out and interpret their needs in the search for Western technologies. Second, having been isolated from international developments, the Soviets have not standardized many of their industrial approaches in accordance with worldwide trends. Soviet organizations have difficulty in maintaining an adequate store of supplies and spare parts to carry out development and production using their own technologies, let alone foreign technologies.

Many American scientists are tinkerers. Once they have a new idea, they adjust their experimental equipment and develop their data systems to try out the idea and to set the stage for further use of the idea. Many American engineers are pragmatists; the test of a new design is not whether a new system can be installed but whether the system operates as advertised over a period of time. We call this type of concern follow-through engineering. These traits of our scientists and our engineers ease the transition from research to practice and create products with integrity. These traits are in scarce supply in the USSR.

Technology transfer difficulties are often overcome in the Soviet military sector through the brute force approach. Not being excessively concerned with costs and usually having requirements for only limited production runs, Soviet military organizations have been able to assign many specialists and large financial resources to big projects. In time, they figure out how to use or reproduce the foreign technologies even if it means handcrafting each product and spare part. Often they use only certain aspects of the newly acquired technologies to upgrade their own approaches. In other cases when they rely totally on copying the foreign achievements, they probably could have developed comparable technology of their own with the investment of fewer resources. However, many Soviets, including Soviet design engineers, are skeptical about the quality of advanced technologies developed in the USSR in comparison with similar technologies developed in the West.

*   *   *

With regard to science, the benefits to the Soviet military effort of exposure of Soviet scientists to Western approaches will not be short-term. As we have seen, the advancement of science is largely an educational process: scientists learn from their own activities and from the approaches and results of their colleagues. The higher the quality of their colleagues and the more developed the research environment, the richer the learning experience. Individually and collectively, scientists apply this knowledge. New discoveries and ideas evolve and are then gradually translated into useful products and services. These applications are sometimes the result of highly orchestrated R & D activities and sometimes the result of serendipity.

Scientific cooperation between the US and USSR contributes to the educational process for participants from both countries. Occasionally, the participating scientists see specific achievements that have near-term relevance for the defense authorities in both countries. Such exposure could influence research priorities and methodologies which eventually benefit military efforts. But the likelihood of near-term military payoff in science is quite remote given the distance from the laboratory to the factory.

During my tenure as the director of a US Government laboratory in the early 1980s, I led many attempts to transfer scientific techniques from our institution to others. For example, we had developed standardized approaches for analyzing chemical pollutants brought from the field into the laboratory, and we were asked to transfer our techniques to other federal and state laboratories so that there would be a uniform approach throughout the country. We had very advanced techniques, the other laboratories needed them to comply with environmental regulations, and we were eager to facilitate the technology transfer. However, the other laboratories were reluctant to give up the techniques that they had used for many years. The personnel in the other laboratories also required hands-on training to use the new techniques, and some of the laboratories had different analytical equipment than our laboratory, which needed to be modified for their use. The process of technology transfer to these laboratories, which was one of our highest priorities, took many months of intensive interactions. We worked intimately with the specialists in the laboratories, prepared extensive documentation, and, not surprisingly, completed the project with high costs.

Drawing from this one not unusual example, how can we assume that casual Soviet visitors to American laboratories will walk away able to transfer techniques they observe back to the USSR? It may happen in rare cases, but we should not underestimate the difficulty of such a task. Scientific exchanges are an educational experience and should be viewed as such rather than as highly targeted technology transfer efforts with predictable results.

*   *   *

Looking ahead, our country has many methods for protecting American science and technology from exploitation by the USSR. Security classification remains the principal means for safeguarding our most sensitive secrets. Export control laws and visa restrictions can be invoked when necessary. Limitations on

US Government scientists and on recipients of government grants and contracts and governmental pressures on our companies and universities will undoubtedly be employed when national security questions arise over proposed contacts with Soviet specialists. Our intelligence agencies will continue their vigilance to deter Soviet efforts to obtain our technology illegally.

But how effectively can we protect our scientific and technological achievements which are considered sensitive, but not so sensitive as to warrant security classification? For how long? What will be the direct and indirect impacts on our own scientists of efforts to deny Soviet access to American discoveries? How severe will be the side effects of our protection efforts in inhibiting access to important information by our scientists and by our friends abroad?

More fundamentally, how important is denial of Soviet access to our achievements in areas where they have relatively uninhibited access to similar achievements in other countries? How important is denial of Soviet access to advancements of the 1990s, when they are having enormous problems—including many problems unrelated to access to technology—in bringing large sectors of their technological base up to our levels of the 1970s?

These questions have been asked in the past. The US Government always tilts toward high levels of protectionism of our science and technology on the grounds that we simply cannot help feed the Soviet military machine. "If we have it and the Soviets want it, it must be better than what they can find elsewhere; and they must be able to use it," say the protectionists.

The US Government can protect some achievements for a limited period of time, including some critical unclassified advances which offer considerable payoff for military systems as well as for commercial production. Control efforts should be directed to a relatively small number of these achievements. The United States should concentrate on preserving technological lead times—lead times for producing new military systems and components which could make a significant difference in the military balance.[17]

Refuseniks, Dissidents, and Scientific Exchanges

No abstract ideas or principles imposed on people
are worth the tears of a single child.
Fyodor Dostoyevsky

During the fall of 1988, Soviet Foreign Minister Eduard Shevardnadze stated at a meeting of UNESCO, "The first commandment of our perestroika is the fundamental value of a human being, his life and dignity, and his social development."[1] No longer can Americans simply dismiss such words as meaningless rhetoric. The changes within the USSR in the field of human rights are profound. They have exceeded even the most optimistic expectations of many human rights activists. By every measure, Soviet citizens enjoyed more freedom during 1988 than at any time since the ascendancy of Stalin more than 60 years ago. These changes are particularly important to the scientific communities throughout the world.

During the past decade a number of prominent American scientists have led the international calls for changes in the Soviet political system which has denied Soviet citizens many basic human rights taken for granted in the West. Some of these American advocates of change are quite influential within American scientific organizations that carry out programs of cooperation with Soviet institutions. They have urged their colleagues to link American willingness to cooperate with Soviet scientific institutions with Soviet performance in the area of human rights.

Most of the concern within the American scientific community can be traced to the plight of a number of Soviet scientists who have been imprisoned for political reasons (often called dissidents) or who have been refused the opportunity to emigrate (the refuseniks). The US Congress and the Executive Branch also share this concern. During each of the four Reagan-Gorbachev summits, discussions of refusenik and dissident scientists were high on the agenda; and a principal theme of Reagan's visit to Moscow in 1988 was human rights.

In the last several years, dramatic changes in official Soviet policies have finally prompted the release of well-known Soviet dissidents and refuseniks. Also, in the spirit of glasnost, public denunciation of acts of terror and oppression of earlier years has been published in the Soviet press and depicted in popular films recently released throughout the country. While the KGB still retains a strong hand in setting the limits to protest and demonstrations, the changes in the candor of the political pronouncements and in the handling of many individual human rights cases cannot be denied.

In May 1988 the Commission on Security and Cooperation in Europe, composed of US senators and congressman, which includes in its mandate the monitoring of human rights performance in the USSR, catalogued Gorbachev's record in this field during 1985–1987 as follows:

- The release (based on written pledges not to engage in further "illegal" activities) of 383 known political prisoners—329 of them freed in 1987—including such renowned activists, allowed or forced to emigrate, as Anatoly Sharansky, Yuri Orlov, Anatoly Koryagin, and Josif Begun.
- The return to Moscow of Nobel laureate Andrei Sakharov after nearly six years of exile in Gorkiy.
- The release of 83 individuals—64 of them in 1987—who had been held in psychiatric hospitals because of their political views and activities.
- Tolerance of sizable public demonstrations by aggrieved minorities—Jews and Crimean Tartars—and aroused nationalities in the Baltic States and Armenia.
- New latitude—to build churches, enroll seminarians, and import or print Bibles—for religious faiths willing to accept official oversight of their activities; and
- A sharp increase—from very low levels—in the numbers of Jews, Armenians, and Germans permitted to emigrate and in the permissions granted Soviet citizens to visit family members in the United States.

The report also noted:

- The 400–1000 citizens still held in prisons and camps for acts of political conscience.
- The 95–150 people being subjected to psychiatric treatment because of their dissenting views.
- The forced emigration of Baltic activists and the resurgence of anti-Semitism in the guise of Russian nationalism.
- The restrictions being imposed not to control, but to preclude, public protests, including a show of military force in Armenia.
- The continuing refusal to allow religious believers and clergy to proselytize—even through Sunday schools.

- The monopoly of mass media by the government and the lack of access by the general public to the foreign press.
- The continuing attacks in the Soviet press on independent political movements such as the independent peace groups.
- The remaining, major obstacles to emigration and travel— including the hardships those barriers work on tens of thousands of still divided families.
- The absence of either legal guarantees or an independent judiciary able to insure justice for Soviet citizens who not only "know" their rights but seek to "act upon" them.[2]

By the fall of 1988, emigration levels had increased dramatically, averaging over 3000 Germans, 1200 Jews, and 1000 Armenians per month—increases of five- to tenfold over the previous several years. The number of political prisoners in psychiatric confinement was believed to have dropped to fewer than 100, continuing a steady downward trend. Of the 11,000 refuseniks whom Reagan had identified for Gorbachev at Reykjavik in 1986, between 8000 to 10,000 had left the country.[3]

*   *   *

Viewing Soviet activities in the area of human rights through Western eyes can be very misleading. Individual rights in communist societies differ markedly from rights in democratic societies. Changes in the obligations and responsibilities of the individual in the USSR will not come easily and may affect only a handful of people at the beginning. Yet the breezes of change have turned into strong winds. Glasnost has let the genie of freedom out of its opaque bottle of the past: the flag of the former independent state of Lithuania was hoisted over the capital of the Soviet republic; Gorbachev admitted to the leaders of the Orthodox Church mistakes of the past and promised a new church-state relationship; and the criminal code will drop the death penalty for economic crimes.

Pronouncements are being made about the rights of the individual at all levels of Soviet society. Some entreaties to work harder and enjoy the rewards of individual efforts are familiar refrains and fall on numbed ears of Soviet workers. New calls for broader participation in the political and economic decisions of the country have somewhat greater attraction. Still other appeals for the bureaucracy and the KGB to move off the front row as the watchdogs of the welfare of the people are applauded but swallowed with a grain of salt.

The response of American scientific institutions to changes in the USSR in the field of human rights is important. From the perspective of the refuseniks, these institutions are a source of strength and hope. At the same time, a few Soviet scientific leaders are now in the forefront in pressing for change, for they realize better than anyone else the importance of acceptance of the USSR at the international tables of science. These leaders of Soviet science who seek a Soviet state more sensitive to the individual need reinforcement for their efforts from the West. The continuing criticism of them by Americans who want more immediate and more drastic changes needs to be balanced with recognition for the progress in resolving cases of Soviet scientists who have been denied basic human rights.

*   *   *

Many books analyze the Soviet constitution and its references to human rights, including the rights to an education, to a job, to housing, to a pension, and to health care. Until recently, American authors had ridiculed the references to freedom of speech, freedom of the press, and freedom of assembly. Now Soviet journalists have taken the lead in underscoring that guaranteed civil liberties include the right to enjoy freedom of scientific, technical, and artistic work; the right to take part in the management of state and public affairs; and the right to associate in public organizations.

These rights are circumscribed in the constitution with such phrases as "in accordance with the interest of the people," "in accordance with the aims of building communism," and "not be to the detriment of the interests of society." Such provisos have frequently been invoked to provide a legal basis for denials of rights. Most importantly, the Communist party is the sole arbiter of significant disagreements over conflicts between the interests of individuals and the interests of the state.[4]

Nevertheless, the presence of concepts of human rights in the constitution and in other legal documents help Soviet reformers find historical roots to justify positions considered to be extreme. In the past a number of the more highly publicized human rights cases were handled without the due process called for in the constitution and in Soviet laws and regulations. In recent years, and particularly now, Soviet authorities have tried to avoid routine perversion of judicial procedures.

The USSR is a signatory of the 1974 Helsinki Accords, which is usually referred to as *the* human rights treaty. They also participate in the continuing international effort to monitor compliance with this international treaty. The Accords have special provisions on fundamental freedoms and on human contacts, including reunification of families separated by political boundaries.[5]

The Accords have provisions on many other aspects of international relations as well. Indeed, the Soviet motivation for signing the agreement probably had little to do with human rights. Rather, the USSR is very sensitive about challenges to the legitimacy of its restructuring of international borders following World War II, and particularly the redrawing of the borders between Poland and East Germany and between Poland and the

USSR. Thus, the Soviets value highly the provisions of the Accords calling for the sanctity of national boundaries which they view as supporting the *de facto* situation in Central Europe. Their signing of the Accords has been interpreted in the West as a willingness on their part to make concessions on human rights in exchange for concessions on the sanctity of boundaries. In addressing its obligations under the Helsinki Accords, the Soviet approach to the provisions on human rights had traditionally been quite defensive. Now they have shifted to the offensive. They expound the humanitarian nature of glasnost and perestroika at every opportunity and confess many sins of the past.

The numerous provisions of the Accords are clustered in "baskets" during international debates on compliance with the agreement. While the United States champions the agreement's human rights provisions, it is given pause by other provisions. For example, a section of the Accords is directed to facilitating scientific exchanges in a variety of fields, including the militarily sensitive field of computer technology. Another section includes discussion of cooperation in industrial technologies which are often subject to American export controls. The USSR obviously finds these sections more attractive than the human rights provisions which challenge traditional practices of communism.

Western anger over human rights abuses in the USSR has attracted the attention of the highest levels of the Soviet leadership for the past several years. Gorbachev and others have convinced themselves that the situation in the USSR is not nearly as bad as portrayed in the West, that it is improving, and that greater Western scrutiny can do no harm and indeed may be useful in modifying Western stereotypes of Soviet behavior. After several years of Soviet urging, the United States joined other countries in supporting a major intergovernmental conference on human rights in the USSR scheduled for 1991, although some US officials remain wary that participants will only see carefully staged glimpses of the real life in the country. However, in the age of glasnost, suppressing expressions of feelings by Soviet citizens will be difficult.

During the 1988 Reagan-Gorbachev summit in Moscow, Reagan used almost every public and private occasion to raise the issue of abuse of human rights in the USSR. He was neither subtle nor indirect as he criticized both general policies and handling of individual cases. His speeches and the associated functions in Moscow were welcomed by dissidents, refuseniks, and a small segment of the Soviet intellectual community. This probably angered Soviet officials, but clearly it conveyed his personal concern on these issues and the priority he attached to human rights in Soviet-American relations. He made headlines around the world.

Yet, the general impact on the Soviet public of Reagan's statements was very limited. The public remembered what the Soviet Government wanted it to remember about the summit, namely, the conclusion of the arms control agreement to eliminate intermediate range missiles. With regard to human rights, many Soviets readily accepted the views on the visit put forth in the Soviet press: Reagan's political statements had little to do with reality; he was both misguided and rude; the United States should clean up its own act toward blacks and American Indians before preaching to the USSR on the rights of Soviet citizens.

*   *   *

Let me return to my personal experiences to help document an important chapter of Soviet history. Prior to 1988 I had always heard that scientific seminars organized by refuseniks in Moscow were small, discreet affairs devoted to discussions of relatively narrow scientific topics. Their primary purpose was to provide an opportunity for scientific interactions among Soviet scientists who had lost their research positions when they declared their intentions to emigrate and applied for exit visas. Through these seminars held in private apartments, they could discuss their research interests with professional colleagues who were similarly isolated from the mainstream of science in the USSR. Western scientists were especially welcomed as participants in the seminars since they brought news and results from the international world of science. The seminars characteristically ended with many requests for the Western guests to carry messages of greetings and desperation back to acquaintances in the United States and Western Europe.

The seminar I attended in January 1988 was a different type of affair. By that time the active refusenik community in Moscow had dwindled to a few hundred; the Soviets had relaxed emigration restrictions. These remaining refuseniks were putting more of their energies into trying to leave the country rather than into maintaining a high level of scientific discussion, as had been the case in the 1970s and early 1980s, when the likelihood of leaving was very low. American visitors had become a common sight at seminars, and discretion had become far less important as the KGB adopted a more tolerant attitude toward the seminars. Direct telephone conversations between refuseniks and friends in the United States had become commonplace: the status of many refusenik cases was well known in the United States on a day-to-day basis, and the need for visiting Americans to report on the condition of the refuseniks to their relatives and friends in the United States had diminished.

On one Sunday I telephoned from our Moscow hotel to the coordinator of the seminars for the Moscow refusenik community and requested that he organize a "seminar" for three representatives of the US National Academy of Sciences for the following Saturday evening, noting that a similar meeting had been held three years earlier. I explained that we were having discussions with the leaders of the Academy of Sciences of the

USSR and that we were concerned over the situation faced by refuseniks. This request hit a very responsive chord. The coordinator immediately agreed to "work us in" between a visiting delegation of US senators and some British visitors.

On Saturday evening we rendezvoused with our host and proceeded to a small apartment jammed with about 50 refusenik scientists. Almost all of them were Jewish. Three places were reserved for us at a small table, but we had to step over two camera crews and several reporters from the Western press corps to make our way to the table. The presence of the Western media underscored the importance to the refuseniks of Western support of their emigration efforts.

The refuseniks had two messages for us. We should use our leverage in discussing cooperation with Soviet science officials to help the refusenik scientists and their families receive permission to emigrate. Indeed, in the absence of formal commitments by the Soviet authorities to take concrete steps in this regard, we should not agree to cooperative scientific ventures. Second, we should insist that the Soviet scientific authorities redefine "state secrets." The official reason for denying most emigration requests by scientists was Soviet concern that state security would be undermined should those scientists who had had access to state secrets be allowed to travel to the West. A state secret was being defined so broadly that it appeared that almost any scientist could be precluded from emigration.

At that time, Western emigré groups estimated that hundreds of thousands of Soviet citizens were trying to emigrate; the refuseniks agreed with official Soviet statements that these estimates were greatly exaggerated. The refuseniks estimated that approximately 10,000 would leave immediately if given the opportunity; however, their estimate turned out to be low, since within the next year the number of Jewish departees alone exceeded 20,000. Of course this number was far less than the more than 200,000 Soviets who were allowed to emigrate in the 1970s. The refuseniks also confirmed official Soviet statements that only several hundred refuseniks had up-to-date and active applications for emigration currently in various stages of the Soviet administrative review mechanism. They underscored that many more were eager to leave; some had been turned down in the past, and these unsuccessful applicants considered another updating of their applications a waste of time. Others did not want to risk losing their jobs by declaring their emigration intentions.

In the days that followed, our delegation had many private discussions with the leaders of the Academy of Sciences of the USSR concerning the refuseniks, and particularly those refuseniks who had previously served on the scientific staffs of the institutes of the academy. We did not directly link specific cooperative projects with Soviet decisions on refusenik cases. However, our Soviet colleagues clearly understood that our attitude toward scientific cooperation was significantly influenced by Soviet progress in the field of human rights, including fair resolution of refusenik cases.

With remarkable candor, the president of the Soviet Academy told the Soviet and Western press following our formal meetings that he was determined to resolve those human rights cases involving former Academy employees as quickly as possible. For the first time, a senior Soviet official publicly acknowledged that the USSR had made many mistakes in handling such cases in the past. He noted that greater recognition of the rights of individuals was consistent with the new policies of Gorbachev. We were witnessing the first crack in the Soviet historical stonewalling on this issue. Within a few months such statements had become commonplace among Soviet officials. Later during our visit, the Academy president's close associates confided to us that they would like to see all the refuseniks given permission to leave. In their view, the refuseniks were not contributing to science; and the debating of their cases was simply diverting the time of many important Soviet scientific leaders away from science.

The Academy of Sciences is but one of the Soviet institutions involved in reviewing refusenik cases. The Ministry of the Interior (which is the home of the KGB), the Ministry of Justice, and the Ministry of Foreign Affairs usually have far stronger voices in the deliberations. Also, the Academy's official involvement is limited to those cases of former employees. During a visit to the United States in late 1988 a senior official of the Soviet Academy reported that only eight cases of denials of emigration for former employees remained. He stated that all of these cases were based on access to state secrets many years ago, and he was confident that some of the cases would be resolved in favor of the refuseniks. When presented with a list of more than 150 refusenik denials by a group of American scientists in New York, he argued that the Soviet Academy simply had no sway in most of these cases since the refuseniks were not former employees.

It is difficult to believe that so many refuseniks had access to such sensitive information at some time during their careers that they were still considered security risks. However, in the USSR, state secrets have always extended far beyond the military security definitions used in the West. Information concerning diplomacy, economic affairs, and industry has traditionally been tightly controlled in the USSR, and these controls continue to restrict information which may be two or three decades old. The Soviet authorities have often been as much concerned about not exposing Soviet weaknesses and shortcomings as they have been about revealing information concerning advanced capabilities.

Now, the rewriting of the events of the Stalin, Khrushchev, and Brezhnev eras by Soviet historians is adding further confusion to the past concepts of state secrets. For example, the Soviet leaders probably ask whether the Soviet role in espionage in the United States following World War II should be unveiled, including those communist activities which were the target of the McCarthy hearings of the early 1950s. Should the early agreements between the USSR and China which led to the development of Chinese nuclear weapons be critiqued? Fragments of these and other intriguing stories of the past are in the heads of many Soviet citizens who have a mind-set dominated by state secrecy. If, as suggested by Soviet officials, the Soviet Government defines state secrets to exclude information older than 10 to 15 years, we may have the opportunity to gain fascinating historical insights that have eluded the West for many years.[6]

*   *   *

The misuse of psychiatry for political purposes has been a particularly repugnant aspect of many authoritarian regimes, and psychiatric abuse has a long history in the USSR. During the Stalin era, large numbers of Soviets who were arrested on many charges feigned mental illness, preferring confinement in mental institutions to likely death in the gulags. Khrushchev then released millions who had been sent to the concentration camps for criminal and for political activities, and he repeatedly claimed that no longer were there political prisoners in the USSR. As is well known, psychiatric institutions were thereupon widely used for confinement of political dissidents. During the 1960s and 1970s, the number of these institutions increased significantly as confinement could be ordered by individual Soviet doctors without the opportunity for judicial review of such orders.

My insights in this area have come primarily through the reports of former inmates of Soviet mental institutions as published in the Western press over the years. However, my one exposure to a Moscow psychiatric ward in 1965 left me with great apprehension concerning the ease of manipulating psychiatric care for political purposes. While serving at the US Embassy in Moscow, I visited an American scientist who was temporarily confined to a psychiatric hospital which was often described as more of a jail than a hospital. As I sat in the main hall of the facility waiting for him, dozens of inmates dressed in gray pajamas and carrying soup bowls walked aimlessly past me, just as in the popularized films which have been shown over the years. They all seemed to be in a daze, and I could not help but wonder just why they were there.

The stories of political prisoners being subjected to psychiatric abuse in the USSR fill many novels and history books and reflect the most brutal aspects of the communist system. Convoluted Soviet psychiatric theories such as "sluggish schizophrenia" to justify unorthodox and reprehensible practices have been the subject of international ridicule.[7] International psychiatric associations have repeatedly condemned Soviet practices and have not recognized Soviet specialists as legitimate members of the international scientific establishment. Specifically, in 1977 the World Psychiatric Association adopted a resolution aimed at unacceptable Soviet practices; and in 1983 when faced with certain expulsion from the Association, the Soviet All-Union Scientific Society of Neuropathologists and Psychiatrists withdrew from membership.[8]

Glasnost has focused a spotlight on abuses of psychiatry within the USSR. In recent months many articles in the Soviet press have criticized the mismanagement, brutality, and lack of professionalism of the mental institutions. Psychiatrists have been accused of placing troublesome relatives in institutions. Indeed the whole subject of corruption within the psychiatric establishment has been a lively topic for the media.[9] Still, the Soviet Government is reluctant to acknowledge that in the past psychiatry was used for political purposes, an acknowledgment being demanded by some Western scientists prior to acceptance of Soviet organizations as legitimate members of the international scientific community.

Meanwhile, there has been a steady stream of political dissidents being released from Soviet institutions to emigrate to the West or to resume life with their families. Some of these bring with them firsthand experiences with psychological and chemical measures used to quell their spirits and to modify their attitudes. Others bring secondhand stories of what has happened in mental institutions over the years.

During 1988, a new law was passed to protect citizens from arbitrary commitment to Soviet hospitals, indicating a recognition on the part of the Soviet leadership that no Soviet institution is beyond external scrutiny. While the implementing regulations are still being drafted, two important steps are to be taken. First, relatives of committed patients will have the right to appeal the decisions of doctors. Second, the institutions for the criminally insane will be transferred from the jurisdiction of the Ministry of Interior to the Ministry of Health. Is this really a handoff from the police to legitimate doctors? Let us hope so.

*   *   *

For many years the United States Government has attempted to apply pressure on the Soviets to change their ways in the field of human rights. Our government has tied trade policies to Soviet restrictions on emigration and has initiated and cancelled cooperative scientific programs in response to political repression of Soviet activists. American scholars disagree, however, as to whether these policies have had a significant impact in forcing the Soviet Union to liberalize its policies toward repression and emigration.

With regard to trade, the Jackson-Vanik legislation of the mid-1970s links our trade policies very directly to emigration policies of the USSR and Eastern Europe. As to the effect on the USSR, a persuasive article in the 1986 winter issue of *Foreign Affairs* argues that the pressure has not worked, since Jewish emigration rates from the USSR dropped significantly immediately after the legislation was passed. The Soviets clearly tightened emigration policies, most likely in response to this American policy which they considered flagrant interference in their internal affairs. In the late 1970s, emigration rates increased as prospects for arms control improved. Soviet officials then sought a waiver of Jackson-Vanik, only to be rebuffed as different views developed between the president and Congress. Emigration rates thereupon again fell. While our law was intended to help potential emigrants, it probably hurt them, at least in the short run.[10]

As previously noted, the Soviets are interested in expanding trade with the United States; but they have many options in seeking other trade partners. Trade between our two countries is currently only several percent of overall Soviet trade. Thus, in the case of the Jackson-Vanik legislation, the trade incentive for liberalization of emigration policies is not sufficient to offset the international and domestic embarrassment to the USSR of appearing to knuckle under to American attempts to dictate its internal policy.

Supporters of Jackson-Vanik will not concede that the legislation may have been counterproductive by delaying emigration while opening trade doors for others. Regardless of the impact of the law, these supporters are very pleased to have it on the books, since they can identify with a very visible effort designed to help oppressed populations in the USSR. They believe strongly that the law is a political statement which is important for the world to hear. The moral tone of their arguments often drowns out logic within the Washington political milieu, and it is likely that the legislation will remain in place for the foreseeable future.

Jimmy Carter considered himself the first human rights president, and he took his campaign into the international arena with vigor. Special offices and programs were set up within the Department of State and throughout the world. Human rights officers were positioned in the US Embassy in Moscow to keep in touch with politically disenfranchised Soviet citizens and to report on Soviet violations of international norms. Soviet scientists have been among the most important contacts for these human rights officers.

Ronald Reagan did not hesitate to continue the efforts of his predecessor, with particular attention riveted on exposing the indignities of the "evil empire." Reagan was known for his concern over the plight of individuals, and he was emotionally moved many times by the reports brought to his attention of highly personalized examples of Soviet oppression. He strongly supported the efforts of the Department of State to ensure that all important discussions with the USSR included a heavy emphasis on human rights. Thus, there was an implicit if not explicit linkage of all American policies, including policies on scientific exchanges, with Soviet performance in the field of human rights.

On several occasions I met with senior White House advisors to former President Reagan to present the case for greater attention to the opportunities for the United States to benefit from scientific cooperation with the USSR. Each time I received the same message. After the Soviets released some of the key dissident scientists, the White House aides would support my views during the administration's discussions of scientific cooperation.

Private American institutions conducting exchange programs have widely varying views on whether to link human rights and scientific exchanges. Most exchange organizations are not interested in becoming involved in the human rights debates which inevitably complicate the administration of exchanges. At the other extreme, during the 1980s, some American scientists called for a total suspension of bilateral scientific interactions pending resolution of a number of human rights cases of Soviet scientists.

In 1980 the National Academy of Sciences, in response to the exile to Gorkiy of Foreign Associate Member Andrei Sakharov and to growing concerns over Soviet actions in Afghanistan, suspended its program of bilateral workshops with the Soviet Academy of Sciences and did not negotiate a new interacademy exchange agreement when the existing one lapsed later in the year. The two Academies continued the exchanges of individual scientists on a less formal basis, with American scientists making their own decisions whether to participate. The National Academy's subsequent efforts to encourage Soviet scientific leaders to intervene in human rights cases were greatly hampered, since communications with these leaders rapidly atrophied with the suspension of the formal relationship. At the same time, the Academy continued supporting the interests of its members in direct appeals to the Soviet political leaders to intervene.

After intensive debates among the members of the National Academy as to whether the institution should again have a formal exchange program and the attendant communication channels or if it should continue to refrain from such a relationship in view of Soviet actions related to human rights, the Academy took the initiative in early 1985 to have the interacademy scientific exchange program formally reestablished, which was finally accomplished in 1986. At that time, a new provision was included within the agreement calling for regular interacademy discussions of the environment affecting cooperation, a provision which has been clearly understood by both sides to include consideration of human rights issues.

*   *   *

At the top of the list of recent advances in promoting the basic human rights of individuals throughout the world are the many contributions of Andrei Sakharov, one of the fathers of nuclear fusion in the USSR. Sakharov has clearly become a major figure in support of perestroika.

Sakharov was elected a Foreign Associate Member of the National Academy of Sciences in 1972, but he was not permitted to travel to Washington to accept this honor for many years. Finally, Sakharov signed the membership book in the Great Hall of the National Academy of Sciences in November 1988. As I watched this event, I was convinced that he was the most iron-willed individual I had ever encountered. Through many years of persecution, he never wavered from his steadfast dedication to championing the causes of human rights.

Ten months earlier I had met Sakharov in Moscow. His convictions were the same, but his appearance and his outlook were remarkably different. At a Soviet reception in Moscow in January 1988, he had stood in a corner by himself, largely ignored by his Soviet colleagues in attendance. He looked frail and was uncertain of his future. In Washington he appeared robust and confident. In the interim he had spoken out at press conferences in Moscow, had become a source of advice for Gorbachev, and had been selected to serve on official committees within the USSR and internationally. He had been elected by his peers to membership on the prestigious Presidium of the Soviet Academy of Sciences. Sakharov is truly a remarkable champion of freedom and an extraordinary man.

*   *   *

In order to protect the interest of the State, Soviet policies which limit human rights have always been firmly embedded in the fabric of communist principles. Many of these policies are now being directly challenged within the USSR, not only by political dissidents but also by important intellectual leaders. They are being challenged indirectly through rumblings within the ethnic communities, particularly in the Baltic states and in the Caucasus, and through the religious communities which have survived many decades of hardship.

Still, the majority of the Soviet population has experienced only the communist approaches of the past: alternatives are unknown. Protests by small groups of dissidents in Moscow, demonstrations in Armenia, and complaints by religious activists used to be considered irritants and received little attention or support among the general population. But the situation has changed, at least in the eyes of the government. Now when the local government of Estonia calls for economic independence, as it did in late 1988, Moscow pays attention.[11]

Nonetheless, while the general Soviet population cares passionately about working conditions, consumer goods, and recreational opportunities, they don't think about the Jewish minority, and they don't want to know about the conditions of mental institutions. Freedom of speech is not a major issue because even in the era of glasnost, they are not interested in making trouble. They have become accustomed to the suppressive tactics of the KGB, and they have little incentive to directly challenge this type of authority.

The population, however, is genuinely concerned about the prevention of war. They fully support the Soviet calls for a total elimination of nuclear weapons. They respond enthusiastically to the Soviet contention that the most basic human right is the right to survive and that this right must receive the top priority of the world community.

Meanwhile, deepening economic deprivation is leading to greed throughout the country as the population increasingly tries to share in the benefits from the underground economy and from all types of schemes to profit at the expense of the state. Soviets are consumed with their economic rights, rights that are increasingly linked to internal battles for larger pieces of a pie that is baked by an inefficient economic system.

The American role model for human rights doesn't sell to the general public in the USSR. Soviet perceptions of crime, poverty, and unemployment throughout the United States raise many doubts about the virtues of American democracy. At the same time, among those Soviet citizens who are enjoying a comfortable lifestyle, thoughts increasingly turn to some of the advantages of the West. Thus, it is not surprising that the well-established academicians are the leading advocates of social and political reform, rather than the struggling young scientists and intellectuals.

As the Soviet leadership attempts to emulate Western technological successes, it has quickly realized that the ingredients of Western approaches include more than management techniques and technical skills. Western approaches work in Western societies which have entirely different standards of value and behavior. If they are to work in the USSR, adjustments in the role of individual initiative as well as in the role of organizations are critical.

Finally, with regard to Soviet-American cooperation in science and technology, many unusual opportunities to establish new patterns of cooperation with a wide spectrum of Soviet institutions and scientists are unfolding. In the long run, such interactions will surely help open the windows which allow the currents of genuine freedom to circulate within the USSR. In the short run, linking cooperation with immediate Soviet progress in moving to a more humane state requires great care. The leaders of the Soviet scientific community are trying to become full-fledged members of the team in the USSR that determines policies affecting many aspects of Soviet life, including human rights. They need to be reminded regularly about painful situations in the USSR which they prefer not to see, but the limits of their influence must also be recognized.

CHAPTER 9

Economic and Scientific Decline in Eastern Europe

An unenlightened nation is the enemy of itself.
Bulgarian writer Todor Vlaikov

Will Eastern Europe become Gorbachev's Achilles' heel? Will political turmoil within this strategic and ideological buffer zone with the West force further Soviet military interventions in the region? Will economic problems plaguing obsolete industries throughout the area require Soviet bailouts? Can the Soviets control the region's thirst for modern technologies which will only be satisfied by a drastic reorientation of domestic and foreign policies by both the Soviet Union and the East European countries?[1]

As Gorbachev presses for rapid reforms in the USSR, many disenfranchised factions in the countries just to the west of the Soviet frontier have become increasingly vocal in calling for much more radical changes in Socialist systems, particularly in Eastern Europe. Glasnost encourages the sharpest critics of communism and of central economic planning to speak out.

For the first time, Hungarians are publicly discussing the uprising of the Hungarian Freedom Fighters in 1956 and the role of Soviet tanks in quelling the rebellion. When I was in Budapest during the thirty-second anniversary of the uprising, my Hungarian colleagues were able to attend public meetings to reflect on these events. However, those from the older generation were dismayed that many young Hungarians at the meetings had never before heard of these events and that the youth were reluctant to believe that the bloody battles had indeed taken place. They, like the youth in the USSR, are in the process of learning the grim truth about the suppression that accompanies the Communist transformation of societies.

At about the same time in Prague, the police were dispersing street demonstrators commemorating the twentieth anniversary of the 1968 effort of the former Czechoslovak leaders to introduce wide-ranging political and economic reforms, an effort that also had been thwarted by Soviet tanks. According to Czechoslovak scientists who were eyewitnesses to the 1968 street rallies, the suppression of the demonstrations reflects the very real worries of their aging party leaders—leaders who remember the peace and tranquility of the Brezhnev era. Just 18 months earlier I was in Bratislava on the day of Gorbachev's visit to that city in southeastern Czechoslovakia, where the spontaneous public turnout to welcome him and to herald his turning away from the orthodox precepts of Marxism-Leninism was unprecedented in recent history. However, this spark of popular hope of 1987 has not yet been rekindled.

In Poland, Solidarity remains a symbol of pent-up frustrations of an entire population over the political, economic, and social deprivations that have resulted from adoption of the Communist model. Even the loyal party members whom I encounter in Warsaw recognize that the calls for change symbolized by Solidarity must be heeded. They sense the movement from economic frustrations to political anger among the general populace as living conditions continue to decline and as internal confrontations become more commonplace and more dangerous.

Talking to colleagues in Budapest, Warsaw, and Sofia, I regularly feel the winds of change blowing through the rapidly expanding openings in the Communist facade. Discussions are candid—in the press, on the street, and at the restaurants. In East Berlin and in Prague, the comments about political change are less frequent, more discreet, and less optimistic. Meanwhile, in Bucharest, my acquaintances are so preoccupied with economic problems that they have little incentive to look beyond the next few days and weeks.

If changes in Eastern Europe come too fast, conservative elements in Moscow will surely try to blunt Gorbachev's entire campaign at home and abroad. If changes occur too slowly, the stirrings in Eastern Europe are likely to go out of control. As aptly described by *Newsweek*, there are "cracks in the bloc."[2] Referring to Soviet political control over Eastern Europe as a "historical parenthesis" that has yet to be closed, the *New Yorker* describes the problem as "what to do about the politically obsolescent, intellectually discredited, economically unworkable structure that has linked this part of Europe to the Soviet Union."[3] Former President Carter's national security adviser Zbigniew Brzezinski says, "The people are restless, and the ruling bureaucratic elites are by and large demoralized and fearful."[4]

US policy has consistently been sympathetic toward the oppressed populations of Eastern Europe. Americans have deplored the existence of the Iron Curtain and the subjugation of personal liberties, but the United States has not resorted to force even when Soviet provocations into the region have exceeded the limits of our tolerance. Rather than risk military confrontations, the US Government has attempted to exert diplomatic and economic pressures in support of efforts within the countries to gain greater independence from Moscow and greater personal freedom. We Americans keep hoping that the Soviets will eventually relax their political domination over the region, leading to the evolution of Eastern Europe into a zone where the USSR maintains considerable influence without the need for troops—perhaps along the model of Finland.

The US Government concentrates on using formal diplomatic channels to encourage changes in the policies of the East European Governments. Our government also encourages many American organizations to interact on a continuing basis through exchange programs with individuals and institutions in Eastern Europe.

Again, at the official level, US representatives have been able to influence developments in Eastern Europe through the banking and economic organizations which are concerned with the large foreign debts of several countries of the region. Also, each of the countries wants better access to American markets for its exports, access with minimal import duties. Thus, the United States can exert some pressure in bilateral trade negotiations even though the levels of trade are small.

The growing technological gap between East and West also offers us important opportunities to help effect change in Eastern Europe. Each of the countries of the region is seeking high technologies from the West, including many technologies such as high speed computers which are currently embargoed for export because of their potential use in military systems. Each country would like to have greater opportunities for its scientists to work in American research laboratories, since possibilities for world-class research are limited in Eastern Europe. Finally, most of the countries are seeking financial and technical assistance through the World Bank, UN agencies, and other international organizations. US policies influence all of these activities, and increasingly science and technology are central to such policies.

*   *   *

Changes will not come easily in Eastern Europe. For more than four decades Poland, East Germany, Czechoslovakia, Hungary, and Bulgaria have been squarely under the military, political, and economic domination of the USSR. While Rumania repeatedly demonstrates its determination to develop its own foreign policy and to choose its own trading partners, that country remains clearly within the Soviet sphere of influence geographically, ideologically, and economically. In 1948, Yugoslavia chose its own road to socialism; Yugoslav diplomats repeatedly stop by my office to remind me that we should never lump Yugoslavia with the Soviet satellites when referring to Eastern Europe. Albania, the poorest of the countries of the region, continues its steadfast policy of isolation from the influence of the United States or the USSR.

Soviet military forces have never been far from potential trouble spots in Eastern Europe. As noted above, Soviet tanks promptly responded to the uprising of the Hungarian Freedom Fighters in 1956 and to the liberal manifestations of Czechoslovak reforms in 1968. Most recently, Soviet forces have stood ready to quell demonstrations in Poland should they become out of control. The Soviets maintain modest military garrisons of tens of thousands of troops each in Bulgaria, Czechoslovakia, Hungary, and Poland; but 300,000 troops hover near the East-West border in East Germany. The USSR considers Eastern Europe absolutely critical to its national security interests. While political, social, and economic experimentation is now considered legitimate, the USSR has never indicated any willingness to tolerate moves which could change the fundamental orientation of the region away from the USSR.

Soviet troops try to maintain a low profile in Eastern Europe, although this is difficult in East Germany given the large number of forces and the small size of the country. During a recent visit to East Berlin, I was impressed by the Soviet effort to promote the image of Soviet troops as peacemakers. I obtained tickets to the annual cultural performance of the Red Army units stationed in East Germany for the East German public. While the small German theater was less than one-half full, the singing and dancing of the Soviet troops were equal to the highest quality in Moscow. Needless to say, it was difficult to believe that regular army troops could leap as high as the world-renowned Moyseyev dancers and could sing with a resonance that would earn them the titles of People's Artists of the USSR. I was convinced they were imports from Moscow, but the announcer kept calling them sergeants and corporals. In the very last scene this deception became apparent. The female soldiers all came to attention for the last curtain call. However, they didn't know how to position their feet at a 45-degree angle, as is taught in armies around the world. They placed their feet in a ballerina position—heel to instep—and then I knew they were imports from the Bolshoi Theater.

In addition to its military presence in the region, the USSR seeks East European interdependence with the USSR in many areas of economic importance, including science and technology, as a major objective. Initially, the Soviets extracted reparations from the region for Soviet sacrifices in ousting the Germans. A particularly contentious issue was the export of coal from Poland to the USSR for very low prices. Then continued heavy East European reliance on the USSR for materials and resources, and particularly energy resources, became a key factor in the development of bilateral relations between the USSR and its Western neighbors. Recently, the Soviets have insisted on receiving higher quality manufactured goods from the region, which has traditionally exported its best products to the West; reflecting this concern, the USSR is promoting an expanding number of trade agreements and joint ventures with East European partners. The emphasis on joint ventures is also probably intended to help counter East European interest in commercial arrangements with the West.

Against this background of Soviet domination, the East European countries have in large measure patterned their political, social, and economic institutions after the Soviet approach. The Communist parties of the countries provide the central link with the USSR, and to a lesser extent, linkages among themselves. Meetings in Moscow of party representatives provide forums for promoting political hegemony within the region. The internal security agencies also help through their own well-developed back channels to ensure consistency of purpose and to prevent approaches which could threaten the basic directions which have been accepted by the USSR.

As noted earlier, the region can only be described as politically unstable. New uncertainties over successors to the geriatric leaders of the region and over the economic reforms being actively pursued both in Moscow and in some East European capitals now dominate discussions of the future. As the economic situation deteriorates, as environmental pollution becomes a serious threat to the health of the populations, and as the technological lag with the West becomes ever more apparent, internal pressures for change are reaching an all-time high.

Publicly, the leaders of all the countries of the region exude confidence in the general character of the existing political and economic systems. However, the hollowness of this confidence is apparent to even the most loyal party members throughout the region, as rationing is extended to bread in Rumania and to meat and other products in Poland, as the wait for apartments and cars continues to be many years in some of the countries, and as the ever-expanding black market operations in all of the countries continue to denigrate the importance of local currency. Even in Hungary, the steady economic growth of the 1970s has encountered the stagnation of the 1980s; prices are on the rise, and a personal income tax has been instituted for the first time.

Questions are increasingly heard at all levels in Eastern Europe as to flaws in the socialist approach, flaws that are leaving the countries behind their West European neighbors. Memories of the days before World War II are fading, but many still remember when the East European capitals were favorite weekend retreats for the most sophisticated travelers of Western Europe.

*   *   *

In embracing Gorbachev's spirit of reform, the leaders of Eastern Europe tilt his concepts to support their own approaches to reviving their economies. Gorbachev tries hard to bolster their confidence that the widening of the economic and technological gap with the West can be abated. There is little evidence, however, that the new Soviet economic approaches are being used as a model in any other country.

The Hungarians have tinkered with their own private sector experiments and with joint ventures with the West for decades; they claim they invented the concept of economic reforms. The Bulgarians boast of their moves in the last several years to promote entrepreneurship through many incentive schemes. The East Germans concentrate much of their production capacity in large industrial conglomerates; they point to their relatively favorable international financial situation and urge others, including the Soviets, to adopt the East German model. The Poles continue to search for divisions of economic authority and responsibility which will increase production without contributing to additional internal political dissent. The liberal thinkers in Czechoslovakia hope Gorbachev's policies will encourage devolution of microeconomic decisions away from Prague, but they are hard-pressed to find grounds for optimism.

The realists in the region agree that reforms of some type are desperately needed, for the economic situation is indeed bleak. The debt to Western banks seems out of control in Poland, Rumania, Hungary, and Yugoslavia. Interest payments, let alone payments of principal, are leading to consumer shortages of all kinds. Much of the conversation at social gatherings in the capitals revolves around future international meetings to reschedule debt repayments and the hopes for benevolence of international financial institutions. On the street, discussions focus on shortages and inflation. To many, the most highly prized possession is now a rich relative in the West who sends gifts to Eastern Europe and supports family members visiting the West.

Rumainia's apparent success in paying off its debt of billions of dollars to the West has turned the country of 22 million persons into a huge island of despair. Survival becomes the top priority for a large segment of the population. As a foreigner in Rumania, I am regularly served meat; however, I feel uncomfortable knowing that this luxury is not available to the population. Let us hope that the future benefits of economic independence, free of the shackles of debt, together with social reforms will warrant the current sacrifices of this heroic population.

Meanwhile, the people of Poland, Yugoslavia, and Hungary long for the good old days of decades past. They now understand how they were able to enjoy a rising standard of living during the 1970s without working too hard—namely, their lifestyles were buoyed by foreign loans, loans with repayment schedules which are much more difficult to meet than had been anticipated.

In responding to pressure for Soviet access to the higher-quality goods, the countries must abandon some of their export markets in the West, markets that will not easily be regained. Also, the developing countries of Asia and the agricultural countries of Europe, such as Spain, have moved into some of the traditional Western markets for East European goods, particularly in the fields of textiles and agriculture. As the East European countries attempt to penetrate markets for manufactured products in the Middle East and Asia, they encounter increasing competition from Japan and Korea, which achieve low costs of production while providing higher-quality goods.

In serious straits to pay for their import requirements, the countries are adopting many approaches to obtaining foreign currency. They widely recruit foreign students who are willing to pay modest fees in hard currency to study at their universities and medical schools where special courses are offered in the English language. Americans who cannot gain admission to American schools, who cannot afford the costs in the United States, or who have strong ethnic roots in the region are responding. The countries offer a wider array of very low cost vacations for Western tourists than ever before. West German tourists line the Bulgarian beaches, sail through the Danube delta in Rumania, and visit the castles in Hungary throughout the summer. In winter they tackle the ski slopes of Czechoslovakia, Rumania, and Bulgaria. The banks of Eastern Europe are even buying up foreign currency at black market exchange rates, although the hustlers still offer better rates outside the hotels of Warsaw.

Increased industrial investments, financed by credits whenever possible, and an influx of technology are being called upon to reverse the economic decline. As has been shown in countries throughout the world, however, finance and modern technologies will do little to improve productivity without comparable attention to the human factors. In all of the Socialist countries of Eastern Europe, the habits of the work forces have deteriorated.

Many workers, having received no rewards for good work, have simply lost their motivation and have slowed down.

*   *   *

Related to the continued reliance on antiquated industrial facilities are many environmental problems; retrofitting old facilities with pollution abatement equipment is simply too expensive. The use of low-grade coal throughout the region and the inadequacy of sewage treatment facilities in expanding large metropolitan areas also contribute to the degradation of air and water resources. These long-standing concerns have become more sharply focused in a new wave of public debates on environmental issues which follows in the wake of glasnost sweeping the region.

I have spent many hours watching Czechoslovak and Hungarian scientific leaders wringing their hands in despair as their governments ignore their recommendations not to construct two dams on the Danube River on their border not far from Vienna. Wetlands will be destroyed, groundwater resources threatened, and fertile lands flooded so that small quantities of electricity can be provided to Austria. I have toured the forests of southern Poland, which are being decimated by pollutants from steel mills, chemical plants, and power plants in Poland, Czechoslovakia, and East Germany. I have seen schoolchildren from northern Bohemia spending their annual one-month "airing-out" periods in southern Czechoslovakia, far from the high-pollution levels where they live the rest of the year. We have heard firsthand reports of the slow destruction of the ecology of the Danube delta in Rumania, where marshland is being squeezed by agricultural development. The environmental problems are very serious.

Grass roots concern over environmental problems is a growing phenomenon in Eastern Europe. The Polish Ecology

Club has spoken out for several years and claims credit for the closing of an aluminum plant. Ecologically minded citizens meet regularly in the churches of East Berlin. The concern generated by the Chernobyl accident and the anger over the destruction of the forests have stirred up young people who have eagerly sought a rallying point for exerting political influence.

$$*\qquad*\qquad*$$

Prior to considering the science scene throughout Eastern Europe, a few words about the historical context of science in the region will help explain the paradox of economic deprivation on the one hand and pockets of good science on the other.

For several centuries scientists have stood tall amidst the chaos that has characterized much of the history of Eastern Europe. Most recently many of the great scientific centers within the universities of the region have survived both the massive destruction of World War II and the subsequent political and economic upheavals. In remarkable displays of courage and commitment to education, many faculty members have ensured that high educational standards are maintained despite political meddling in university activities and stringent budgetary constraints. The university legacy of excellence is one of the principal reasons that the countries of Central Europe remain internationally competitive in a few areas of basic science.

In Poland, Nicolaus Copernicus remains the highly visible symbol of that country's contributions to science. The geodesy and astronomy exhibit of the Copernicus museum is a favorite stop for important international visitors to the Jagellonian University in Krakow. Mendel Square in Brno in Czechoslovakia represents the site where many basic tenets of modern genetics were developed. Zeiss and his colleagues from Jena in East Germany are widely hailed as the early pioneers in the field of optics.

Hungarians frequently talk about the contributions to early development of the US nuclear program by three emigrés—Edward Teller, Leo Szilard, and Eugene Wigner. East Germany claims credit for the early honing of Albert Einstein's talents, and the Poles are proud of the ethnic roots of Marie Curie. Sigmund Freud was born in Bohemia (Czechoslovakia). Today thousands of practicing scientists are included among the millions of emigrés from Eastern Europe who have established permanent residences in the United States. Many of these first and second generation immigrants are the sons and daughters of scientists who had achieved wide recognition in Central Europe.

The Balkan countries of Rumania, Yugoslavia, and Bulgaria do not have as long a heritage of scientific accomplishments as the countries to the north. However, a number of achievements during the past century—ranging from breakthroughs in theoretical mathematics to design of laminated skis—have brought pride and recognition to the area. Many important achievements came from the universities. While the leading scientists of the Balkan region have not become household names in the West, their contributions to the history and cultural life of their countries have been profound.

To the general populations of Eastern Europe, the title of "professor" has always commanded considerable respect. Even in the times of economic despair within the universities, the title has remained a symbol of the importance of education, both in science and the humanities, throughout the intellectual community. Names of early professors adorn buildings, streets, institutions, and medallions throughout the region.

From their earliest days, many of the universities of the region have had an international orientation; and in recent years they have struggled to remain windows on the world. Some of these universities have been both pillars of scientific stability and centers for revolutionary ideas, even as armies moved back and forth and national boundaries shifted. All too often faculties and students were subjected to incredible hardships for carrying on academic traditions, but the institutions rebounded. In Krakow, for example, I heard the story of the mass deportation to concentration camps of 60 leading university intellectuals during World War II after being lured to an "academic" meeting by the Germans. Buildings and facilities of many universities have been ravaged and books have been in short supply, but somehow intellectual zeal has not died. Outstanding students and teachers continue to be attracted to these centers of learning.

The people of Eastern Europe take pride in their scientific heritage, and references to achievements of the past are frequently put forward in efforts to compensate for the scientific weaknesses of the present. For example, Czechoslovak scientists developed the soft contact lens more than 20 years ago, and since that time this achievement has been used to exemplify the potential of science in that country. Unfortunately, there has yet to be an equally impressive achievement which could be considered a successor to the contact lens. Looking back further, the stark contrast between the leadership position of the universities of the past and the current lag in science in comparison with the West has deeply troubled and embarrassed many scientific leaders of Eastern Europe. However, they are now beyond the stage of embarrassment. In a sense, they unabashedly seek to trade their past contributions to science for sympathy and help from the West in overcoming the current financial barriers to scientific research.

In science, the primary intellectual ties have always been with the West, and even the large exchange programs for young scientists between the East European countries and the USSR have not broken these bonds. The East European desire to interact with the West is growing each year, as they sponsor international meetings as drawing cards for busy Western scientists. Even in East German universities where isolation from the West has been the greatest, Western scientists are greeted with genuine enthusiasm; and a trip to the West is the ultimate scientific reward.

*   *   *

Let us turn more directly to the present approach to science in the region. Following World War II, the countries of Eastern Europe adopted much of the Soviet organizational model for science. Large academies of sciences were established in the capital cities, with many research institutions, both old and new, placed subordinate to the academies. In some cases these new academies absorbed the facilities and responsibilities of earlier academies with different orientations. The idea was to concentrate large numbers of scientists and sizable budgets for work on related problems at single locations. This was the Soviet approach. In contrast, dispersal, rather than concentration of research, has generally been the American approach in basic science, except in fields such as manned space programs and fusion research which require large experimental facilities.

The best researchers in each country soon populated academy institutes, drawn by good facilities and good salaries. Leading scientists from universities and industry as well as scientists from the institutes of the academies were elected members of the academies. Through this broadly based membership, coordination of activities among various organizations was to be assured.

After 40 years, the academies of sciences are firmly in place in each country. In Rumania and Yugoslavia, however, the academies do not have large research institutes. In these countries the large research centers are firmly under the control of governmental ministries. In view of the pride in ethnic traditions, provincial academies were established or retained in each of the eight republics and autonomous areas of Yugoslavia and in the Slovak capital of Bratislava, Czechoslovakia.

In general, the academies of sciences are now the leading research organizations in Eastern Europe in most areas of basic research. They command considerable influence within both government and party circles, through their institutional activities and through the stature of individual members. They are clearly the centerpieces of the region's approach to modern science.[5]

Hundreds of "world-class" scientists work in academy institutes, and a smaller number of excellent scientists are affiliated with the universities of the region. These researchers of international stature tend to concentrate their efforts in fields of historical strengths of the countries such as mathematics, the biomedical sciences, and organic chemistry and in other fields not requiring large research facilities.

Meanwhile, the importance of a technical education has increased in recent years, with large engineering and technical schools populating the region. For example, the Polytechnic Institute in Bucharest now has 22,000 students; agricultural schools have been expanded in almost every country; and technician training has become highly standardized at many levels. A large array of specialized research institutions serve almost every branch of industry—institutes for research on individual agricultural crops, on computer systems, on diseases, and on resource exploration.

But what does all this mean? Of what use are the many scientific and engineering achievements of the hundreds of large research institutions in the region if they are not applied? And current economic and technical assessments strongly suggest very limited success in applying modern science and technology to the real problems confronting the region. To be sure,

East Germany has been a pacesetter in optical technology, Yugoslavia has been a world leader in prosthetic devices for artificial arms and legs, and Hungary is the home of the accordion buses. But overall, the levels of technology compared to technology in the West and the economic success of the region on the international market have been on the decline.

A clinical diagnosis of the relationship between education, scientific research, and the standard of living is very difficult. However, the large number of well-trained technical specialists who have little opportunity to use their skills due to lack of facilities or lack of incentives suggests that central planning of science and technology is in trouble in the region. An extreme case is in Rumania, where 20 nuclear reactor engineers are graduating every year, but where it will be many years until the nation's first nuclear plant is in operation. In several countries, and particularly Yugoslavia, the brain drain has reached near-crisis proportions as much more lucrative opportunities are available abroad for the most talented technical people.

*   *   *

As one example of adapting a Socialist system to a changing technological world, Bulgaria, with a population of nine million, provides interesting insights into the problems and challenges of a country closely tied to Moscow. Also, the economic position of Bulgaria has been sufficiently favorable in recent years to permit experiments with new economic mechanisms, and lessons can be learned from these experiments.

Bulgaria was one of the least industrialized European countries at the end of World War II. When I first visited the country in 1958, the living conditions were primitive. I remember meeting a relatively prosperous farmer who had saved for many years to buy a set of false teeth that had become his prized possession.

Today, Bulgaria has an established industrial base, a well-organized agricultural sector, an ambitious nuclear energy program, and a range of export and import activities. The standard of living, while significantly lagging behind Western Europe, has been rising slowly. I have recently visited homes in the cities and in the countryside. While modest, the living quarters are adequate and more comfortable than I had imagined. The norm of 10 people sharing three rooms has faded into history. Also, during the last three years I have seen a steady improvement in the food and consumer items on sale for the general public.

In comparison with most of the other countries of Eastern Europe, Bulgaria is in a favorable position for continued economic growth. Its foreign debt is modest, the expectations of the population are realistic, and the energy and drive of the leaders and the work force are impressive. By Western standards the level of technological development is in its earliest stages. Yet in comparing Bulgarian achievements with other Eastern bloc countries in one important area, namely computers, achievements are quite advanced.

In late 1988 I visited two new Bulgarian factories producing integrated circuit boards for personal computers and for control systems; West European specialists had guided their design and construction. My companion from IBM confirmed that the equipment from Western Europe used on the production lines was state of the art and that the products should be of comparable capability to similar products produced in the West. The Bulgarians were quick to acknowledge quality control problems in earlier factories, but they were convinced that within one year they would have these problems solved in both the old and new plants. Marketing the products in Bulgaria and the Soviet Union, where competition does not exist, is easy, but Bulgarian computers had not yet entered the Western markets. The Bulgarians were skeptical that they could match Asian prices in the immediate future, since they were still dependent on imports of some of the components for their computers.

What explains the modern transformation of this small agricultural country which is little known to most Westerners?

Bulgaria is closely tied to the USSR, both economically and politically. Soviet raw materials available at favorable prices and large Soviet credits have played a critical role in the development of the economy while establishing a level of economic dependency that will not be easily reduced in the near term. Many key Bulgarians were educated in the USSR, and, as in other countries of the region, the Communist party linkages with Moscow are strong. Bulgaria's foreign policy, in particular, closely mirrors the policy of the USSR; but the Soviets have watched rather than dictated the course for managing the economy. While in some countries close ties with the USSR have led to economic decline, in the case of Bulgaria benefits have resulted. When a country is starting from a very primitive base with a rich uncle (the USSR), a centrally planned and controlled economy may be a socially repugnant, but nevertheless economically effective, antidote to poverty.

An important factor in Bulgarian development has been the 30-year tenure of Todor Zhivkov, the head of the party and the head of the state. He has engendered a degree of political stability in this part of Eastern Europe, an area not given to political calm. He encourages innovation and experimentation within limits which are reasonably clear to all. During an extended meeting with him in 1986 and a brief encounter in 1988 he did not hesitate to expound to me and to several other Americans on the importance of science and technology and on the need to give greater attention to individual initiative and individual rewards. The commitment of President Zhivkov and other Bulgarian leaders to using science and technology to accelerate national development is reflected in budgetary allocations, in the orientation of the educational system, and in placards displayed throughout the country heralding the achievements of science. The current movement toward establishing large high-tech state enterprises while also supporting smaller highly innovative firms reflects his quest to enter the modern age.

During the past several years, the government has attempted to move ahead in implementing the concept of decentralized economic planning and management. Among the early items of business has been the development of mechanisms for providing bonuses for successful managers and for finding capital for investments in promising technologies. President Zhivkov believes that these problems can be solved through a variety of economic incentive schemes.

The Bulgarians will continue to consider the Western countries, and particularly the United States, the masters of the innovation process. At least in the near term, they will copy Western approaches to the extent feasible within their own economic system. They will seek joint ventures with Western partners for access to advanced technologies, particularly the fields of robotics, computers, and chemical engineering. However, one of the great pitfalls in the years ahead may be excessive reliance on imported technologies which discourages the development of Bulgarian capabilities.[6]

*   *   *

I have seen scientists wearing heavy coats, hats, and gloves while attempting to carry out precise laboratory experiments in Rumania during the winter. They simply had no heat. I have been unable to visit a number of laboratories in Poland, since they had closed down; electricity was not available, and chemical supplies had run out. In all of the countries, I have admired the ingenuity of scientists who keep imported scientific instrumentation operating without access to needed spare parts and proper maintenance services.

As might be expected, some laboratories are in much better condition than others. The few prosperous laboratories are frequently associated with highly successful manufacturing enterprises or with groups of enterprises or have been successful in generating foreign currency through their research and service activities in cooperation with Western firms. However, only a handful of the laboratories of the region are equipped with experimental instrumentation of the level of sophistication now commonly found in the West.

As in the USSR, moonlighting is commonplace throughout the scientific establishments of Eastern Europe, as scientists seek to augment their base salaries through additional teaching or consulting services or through service on committees which compensate members. Compared with other professions the salaries of scientists are generally good, but like other members of these societies the scientists desire higher incomes. Thus, many laboratories are often unoccupied, with equipment standing idle simply because the responsible scientists are participating in outside projects or jobs.

Also as in the USSR, scientific journals from abroad remain in short supply, and the situation is becoming worse. Locally published journals are available, but in most fields the best East European scientists much prefer to publish abroad, where they can gain international recognition. Due to budget constraints, many scientific libraries of the region must be highly selective in limiting their choices of foreign journals. Much of the current effort is to collect as many issues as possible of reputable foreign journals which are available free of charge through friends abroad and hope that articles of importance to the scientific staffs will be included.

Many Western scientific journals require that authors of articles pay page charges for the privilege of publishing their articles. Since these page charges are in Western currencies, excellent East European scientists who simply have no access to foreign currency do not publish internationally.

Even though travel restrictions seem to be easing for scientists wishing to travel to the West from several of the countries of the region, travel is at the same time increasingly constrained by the lack of funding from the scientific institutes. Even scientists selected to participate in official exchange visits may be required to pay a portion or all of their travel expenses with personal funds. They therefore may not be able to participate.

Opportunities to participate in international meetings within the socialist countries are welcomed by many East European scientists, particularly if Western scientists are in attendance. However, meetings involving only East European and Soviet scientists are sometimes considered more of a diversion than serious scientific seminars. For example, cooperative research projects among the East European scientists in the environmental sciences have been in place for more than 15 years, and numerous meetings have been held on many scientific issues. Yet the East European scientists recognize that these activities are quite sterile without the active participation of Western specialists who have been working in the forefront of some of the areas of interest with far more sophisticated instrumentation and data processing capabilities.

Despite these hardships, the scientists of Eastern Europe carry on. They often concentrate on relatively simple experiments which are important but ignored in the West, where the efforts are directed to more complicated experiments using better instrumentation.

I recently joined a number of leading American scientists, including several members of the National Academy of Sciences, in East Germany and Hungary for meetings on advances in the biological sciences. Many of the results reported by the scientists from these two countries were clearly on the forefront of modern science; such scientists are able to overcome the difficulties of Eastern Europe and to make contributions to the advancement of international science.

*   *   *

East European scientists treasure their lifelines to the West. For most East European scientists these lifelines consist only of occasional letters. However, for the very best young postdoctoral scientists from the best institutions in Eastern Europe, private channels have opened for visits to the United States.

In Hungary, for example, I was surprised to visit one leading institute where the policy is to have one-fourth of the scientific staff working in the West at any given time. Most of these scientists are postdoctoral research assistants at American universities. They receive sufficient funds from these universities to maintain a reasonably comfortable lifestyle in the United States for one to two years. The senior American professors who arrange the visits are generally delighted by the rigorous educational preparation of the scientists in Hungary and by the diligence of the scientists once in the United States, where they cherish every day as a new professional opportunity. A similar philosophy is followed by several laboratories in Poland, although the overseas placements are divided more evenly among the United States and Western Europe. Directors of some Polish research institutes make annual trips to the United States to find places for their new postdoctoral scientists.

Most young East European scientists who come to the United States on such programs return home, provided that they will have future opportunities to travel abroad again and to remain up to date with modern research techniques. There are exceptions, and as previously mentioned, the East European governments are increasingly concerned over brain-drain losses.

With regard to technology linkages to the West, the steady stream of personal computers crossing the East European frontier from the United States, Western Europe, Japan, Hong Kong, and Taiwan vividly underscores the thirst of the East European countries for technologies which will enable them to enter the twenty-first century as developed and not developing countries. The computers come through many channels. They travel on the laps of visitors, in diplomatic baggage, on trucks returning from Vienna, and on ships from the Far East. The East European governments frequently turn their backs on black market operations involving computers, and indeed even government agencies buy computers from the black market.

The East Europeans are under no illusions about their backwardness in high-tech areas, which is symbolized by the current state of their computer technology. Western television programs of all types being received in the region are constant reminders of the technological advances of the West.[7]

$$*\qquad*\qquad*$$

The East Europeans are also sensitive to the low levels of technological achievements of the USSR. Given the choice between a newly developed Soviet computer or a model with half the power developed in the West, they will choose the reliable Western model. The same can be said for Soviet watches that lose time and for Soviet mass spectrometers for analyzing molecules that must be rebuilt.

The lack of reliability of the USSR as a source of advanced technologies is exemplified in their servicing of commercial airliners. The Soviets have historically provided aircraft to Eastern Europe. These countries often have problems obtaining replacement engines and other parts from the USSR. Now for the first time, Poland and East Germany have begun discussions with the Boeing Company and with the West European Airbus manufacturers concerning purchases of commercial planes from the West.

It is not surprising that the East European countries have not embraced with great enthusiasm the Soviet-inspired plans for pooling East European resources in the fields of computers, robots, advanced materials, nuclear power, and biotechnology. Cynical East European scientists have told me and other visiting Americans on more than one occasion that such an approach will ensure that all the countries go down together, since they are all so far behind the West. Nevertheless, several years ago Gorbachev made this integration of the scientific and technological capabilities of the East European countries a centerpiece of his policy toward Eastern Europe. However, many of the ambitious research and development projects, which usually call for control by the USSR, have been largely ignored by the East Europeans who see the real technological future in the West.

*   *   *

The East Europeans have great pride in their stamina to survive the turmoils of history which have with considerable regularity disturbed the region. As I stood with a Polish scientist on the edge of Krakow directly in the pollution plume from the Lenin steel mill in Nowa Huta, I lamented over the deterioration of the quality of life in this beautiful city. "We have survived such troubles for many centuries, and we will continue to survive," was his unvanquished message.

East European scientists have no doubts about their intellectual capacity to address current problems nor their ability to respond to the challenges of new technologies. "All we need is

money" is a common phrase. Most East Europeans are pragmatists. They fully recognize the limitations imposed by the current
painful economic situation, which they increasingly attribute to
the adoption of the Soviet model of central control following
World War II. Economic limitations are reflected in the budgets
of the scientific institutions and in access to foreign currency.
With each passing year the scientists see the international competitiveness of their activities slipping in almost every field.

Meanwhile, the political leaders of the countries tinker with
the governmental structures, issue appeals for greater productivity from the workforce, and make endless overtures to the
West and to international organizations for help. Less and less
are these appeals presented as commercial credits (since they
can't repay their loans) or as mutually beneficial joint ventures
(since they can't attract partners). The East European countries
increasingly seek help the way developing countries seek
help—through foreign assistance.

For many years foreign assistance agencies—particularly
the World Bank and the UN agencies—have been active in
Yugoslavia, Hungary, and Rumania. They have also had limited
involvement in Poland and Bulgaria. However, their early contributions were generally more of a supplement to national efforts than components of doctor-patient relationships that so
often characterize the activities of these agencies in the Third
World. Now, the economic problems have become so serious in
some of the East European countries that foreign assistance has
taken on more of a prescriptive character.

The US Congress has always taken a special humanitarian
interest in Yugoslavia and Poland, dating back to the early post-
war days of food aid for these countries, which are the birth-
places of many million Americans. The humanitarian instincts
of the Congress are again being aroused. Special legislation to
provide food and other types of humanitarian assistance to these

two countries and perhaps to other East European countries on a more sustained basis seems likely. Meanwhile, special Congressional legislation also provides funds to support modest levels of bilateral scientific cooperation with these two countries.

An important outlet for the East European scientists during these times of economic hardship is the network of international scientific organizations which broker international meetings and some international projects. These institutions include UNESCO, the UN Environmental Program, the World Health Organization, other UN science agencies, and a large number of international scientific organizations which operate outside the governmental framework. These organizations have become increasingly sympathetic to the foreign currency difficulties in Eastern Europe and provide support for activities in the region. The East European scientists welcome these connections, which are rationalized as cooperative science and not as foreign assistance relationships.

*   *   *

Turning again to US interests in Eastern Europe, we note that during times of heightened superpower tensions Eastern Europe becomes a strategic factor in East-West relations. Eastern Europe is viewed as the frontline for potential military confrontation. Yet, during times of détente, the United States considers the region in its own right and turns to the attendant opportunities for developing economic, scientific, and cultural relations with the region. Conflict frequently arises between these two perspectives in developing the basis for US policies.

Once again, the US Government has been indecisive as to whether US interests are best served by contributing to economic prosperity within the region or by hoping for economic decline, which could lead to political instabilities and eventual political change. The question of whether East European prosperity conforms to US interests is critical in determining how best to address issues related to the transfer of high technologies which in the long run could have economic impact in the region.

The more than 16 million Americans with ethnic roots in Eastern Europe retain very strong feelings toward the region and often influence US policies, particularly policies which impact on the welfare of the general population of a country. Americans want the people of Eastern Europe to be prosperous and free, but the debate over how to achieve this objective is often sharply divided. Lurking in the background is the specter of feeding the Soviet military machine if we allow technology to flow into the region.

A related aspect of US policy is our long-standing commitment to link US trade concessions to modifications of East European policies in the area of human rights. Indeed, this principle is ingrained in US legislation that permits Most-Favored-Nation status and lower tariffs only for countries which have acceptable track records in the field of human rights. These lower tariffs for imports into the United States require progress in the movement of East European policies away from political persecution and from restrictions on the right to emigrate.

However, trade with Eastern Europe has been at a low ebb, and thus the bargaining power of the United States is not very great. Even in the case of Rumania, which is desperate for export opportunities of any magnitude and which values the political as well as the economic benefits of Most-Favored-Nation status, the stakes are not sufficiently high to extract additional Rumanian human rights concessions. Also, as history has shown, the emigration of Jews from Eastern Europe, which is a cutting-edge issue in the human rights debate, is less influenced by trade relations than by the overall state of relations between the concerned countries and the United States.

The United States has for many years espoused a policy of differentiation among the countries of Eastern Europe. This means that the US Government views each country differently, depending on the steps the country is taking to liberalize its internal political and economic systems as well as on the extent of its autonomy from Moscow as reflected in its foreign policies. Who could argue with such a general principle? The problems come in the implementation of this policy, which constantly runs into the persistent stereotype that the region is a "bloc" of nations controlled by the USSR.

In many ways, US technology transfer policies toward Eastern Europe (with the exception of Yugoslavia, which is not considered a potential adversary), are identical to US policies with regard to the USSR. The basic assumption is that once Eastern Europe has access to a technology, the technology is available to the Soviet military establishment. If only technology acquisition, transfer, and diffusion were so simple.

Hardware that is obtained in the West can of course be shipped from place to place. However, intellectual capabilities are the key to successful absorption and diffusion of technology, and they cannot be easily disseminated. Again, effective technology transfer is a people problem, not simply a hardware problem. There is an enormous difference between having a computer and being able to design and manufacture a computer. There is a great difference between witnessing a scientific experiment and being able to understand and reproduce the experiment. East European scientists and engineers have little opportunity to incorporate modern scientific achievements directly into military systems since they do not design or manufacture modern military hardware. The problems are formidable in effectively transmitting useful information acquired from foreign contacts to Soviet military designers who will use the information effectively.

The US Government restricts contacts between American and East European scientists who are assumed to be channels for flows of high technology to the USSR. Restriction on issuance of US visas to East Europeans scientists and engineers is the principal means being employed. Of course if classified information is involved, such restrictions are important and necessary. However, with regard to scientific exchanges to consider unclassified ventures and joint research, many of these restrictions are accomplishing little more than increasing the barriers to communication. Such communication is important both for scientific and political reasons: East European scientists represent a segment of East European societies that is receptive to new ideas and is in a position to help bring about important changes in attitudes.

From time to time there have been revelations about East European exchange scientists who were in fact professional intelligence agents. Obviously steps are needed to minimize the loss of sensitive information through their activities. However, the frequency and significance of such cases are small, and overreaction to this possibility can be an undesirable barrier to legitimate scientific contacts.

Looking ahead, the United States has several tools for influencing developments in Eastern Europe. On the economic front they include credits, import restrictions, and influence on the policies of international lending agencies and foreign assistance organizations. They also include technology transfer policies that limit sales of hardware or software and policies that impose restrictions on contacts between American and East European scientists and engineers.

In the field of scientific exchanges, funding limitations on both sides are a major barrier to greater contacts. The current two-way flow of scientists could be easily increased in areas of East European strengths of interest to American scientists if more funds were available.[8]

*   *   *

In summary, the pressures for political, economic, and social changes within Eastern Europe are substantial. They come from a restless population, they are stimulated by changes being championed in the USSR, and they are driven by the demands of the world economy.

Realistically, the existing leadership in each of the East European countries will not easily part with the ideas of the past, nor will the newly emerging leaderships be willing to become sacrificial lambs through uncertain political and economic experiments. As Gorbachev is demonstrating every day in the USSR, even modest changes are painful for some, with the benefits not immediately apparent. Changes in Eastern Europe can be extraordinarily important, however, in bringing new economic opportunities to the general population of the countries, which now numbers about 150 million.

Eastern Europe remains of critical importance to Western strategic planners engaged in war games with tanks rolling across the plains of Central Europe. While such scenarios seem remote in the nuclear age, these planners will undoubtedly continue to carry influence in the years ahead and will make a strong case for the importance of the region. Although the East European countries have had very little influence on arms control agreements of the past, their interest in the current discussions of reductions of conventional forces in Europe is intense. The US Government is already engaged in both official and informal dialogues with the leaders of the region to accelerate their education in a field from which they have frequently been excluded in the past—arms control. However, more educational efforts are needed if the East European countries are to become a significant force in arms control negotiations, a force that over time may not always side with the USSR.

In addition to becoming a significant force in the arms control arena, Eastern Europe will remain an important political factor in the development of overall superpower relations. While positive developments in US relations with Eastern Europe often do not seem to have much influence on US-USSR relations, adverse developments in the region, ranging from the smothering of Solidarity to the shooting of US military personnel in East Germany, can certainly retard progress in the development of positive US relations with Moscow.

Scientifically, the region has the potential to remain a locus of research excellence in selected fields. This potential is reflected, for example, in the results of the worldwide competition for highly prized scientific positions in West Germany under the Von Humboldt Foundation. In 1987, Polish scientists held one-third of these coveted positions.

The area of transboundary environmental protection is of great importance for many countries. Eastern Europe is engulfed in environmental problems with substantial international as well as domestic consequences. As the United States pursues its commitment to understand and abate regional and global environmental problems, Eastern Europe deserves very high priority as a partner. Scientifically, there is probably no better region of the world to study the real-world effects of pollutants—effects which are now only predicted on the basis of studies of laboratory animals and greenhouse plants.

CHAPTER 10

Five Years to Change the Course of History

We must succeed. Our country has nowhere to retreat.
Mikhail Gorbachev

According to Robert Kaiser, former Moscow correspondent for the *Washington Post*, we are "witnessing the beginning of the end of the Soviet empire. The fear of Soviet conquest and hegemony that dominated world politics for more than a generation should now dissipate. We have passed the high water mark of Soviet power and influence in the world." He notes that "the facts that made reform necessary describe failure . . . failure is a fact, while the reforms—at least the practical ones affecting the economic life of the country—remain just a hope."

Kaiser then concludes,

> Most of the twentieth-century empires collapsed peacefully, but some only fell under the pressure of violence. The French held on too long, at too great a cost; the Russians may do the same. It is hard now to imagine how Gorbachev or any future

Soviet leader could gracefully yield to the Poles or the Hungarians—not to mention the Armenians or the Estonians—their
independence. But the entire Gorbachev phenomenon was
hard to imagine before it happened. These are amazing times.
The most dramatic political experiment of the century is col-
lapsing before our eyes—slowly, but certainly.[1]

*   *   *

While discussions of the future course of developments in
the USSR become increasingly speculative as political and economic changes take place with unprecedented speed, an alternative to Gorbachev's reforms has yet to be born, and his continued tenure in the immediate future seems likely. However,
Gorbachev himself has vigorously advocated limiting the term
of Soviet officials, including the Secretary General of the Communist Party, to 10 years. Therefore, he will presumably step
down from his position no later than the beginning of 1995, after
one decade of remarkable changes in the USSR.

By that time Gorbachev will be the leader of a Soviet Union
that would not be recognized by his predecessors who left the
scene in the early 1980s. The Lithuanian mouse will have earned
the respect of the Russian bear which had to balance a divided
empire while searching for food. Brains hopefully will have replaced brawn as the calling card of personal stature. Grass roots
initiatives should have achieved wide-scale acceptance. And
perhaps consumer goods rather than guns will be the symbol of
national survival.

On the fast track of political reform, the emergence of genuine dissent and debate within the Communist Party has already begun; and even the evolution of a somewhat independent judiciary may begin. Gorbachev must pay greater attention
to public outcries and demonstrations for better goods and services and for less control from Moscow, especially in the outlying provinces where life is particularly difficult and where ethnic

loyalties run high. If he is to remain in office, he will have to avoid serious internal disorders in central Russia; but the possibility will remain that such disorders triggered by impatience over the lack of tangible benefits of perestroika could provide rallying points for the conservative forces in the upper echelons of the party.

On the slow economic track, Gorbachev will undoubtedly encounter even more difficulties in the effort to double worker productivity by the year 2000, a goal that will still leave Soviet industry hopelessly behind advances of the West. The Western-style advertising which began in 1988 with full page ads in *Izvestiya*, trumpeting the fragrance of French perfumes, will continue to signal the spread of joint ventures with Western firms to help restock the shelves of Soviet stores. Gorbachev will probably resist the pressures of 1989 and not seek imports of refrigerators that chill, jeans that fit, and other immediate necessities to calm a restless population; by 1995 the alternative of encouraging joint ventures may finally lift the consumer sector while strengthening indigenous manufacturing capabilities. In order to subsidize these joint ventures and also to pay for imports to equip industrial facilities in other sectors on a large scale, however, the USSR will have to reach out for new Western credits. Such credits may exceed the limits recommended by Soviet economists and require repayments through larger exports of gold, which could threaten the international price of gold. Meanwhile, Western-made television receivers the size of dinner plates can be expected to infiltrate the USSR and feed satellite broadcasts from the West into Soviet homes during the 1990s, constantly reminding the population of its unsatisfied demand for personal automobiles and VCRs.[2]

Soviet science is changing, as scientists take advantage of new opportunities for pressing their individual ideas, and as debate and disagreement become the rule of the laboratories rather than the exception. New facilities and equipment will no doubt bedeck some research institutes; pay scales should continue to be relatively high; and a new sense of dynamism is already engulfing some institutes which in the past have best been described as sleepy. While the advocates of greater emphasis on basic science now have a powerful voice, the pressure for more immediate returns from investments of scarce resources into research may again become more dominant. A few areas of traditional science strengths such as physics, mathematics, and earth sciences, perhaps buoyed by a new Nobel Prize for a Soviet scientist after a drought of many years, will clearly be protected; but many basic researchers could fall on hard times. Meanwhile, larger research budgets than ever before will surely be targeted on a few areas of applied science in a determined national effort to see tangible progress in the use of biotechnology, automated systems, and more sophisticated electronics. Still, international telephone calls and the use of electronic mail will remain rare adventures for most Soviet scientists isolated from the mainstream of international science.

In the military sphere, Gorbachev will not be watching a world on the way to total elimination of nuclear weapons in the years ahead. We in the United States will never agree; and in the last analysis, neither will he. In 1995 nuclear weaponry will be one of the few vestiges of the Soviet empire of the past and the Soviets' most powerful force in ensuring the respect of the world for the USSR. However, Gorbachev will probably be presiding over significant cuts in the size of the Soviet armed forces and a shifting of economic resources from the military to the civilian sector. By 1995, the United States and the USSR hopefully will have agreed to new limitations on testing warheads and weapons systems. Most importantly, unilateral arms control measures, first by the Soviets in cutting conventional forces in Europe and then reciprocated by the United States in related areas, could become an accepted approach. This route would be an accelerated alternative to the time-consuming formal negotiation process which results in our diplomats curtailing the problems posed by yesterday's technologies rather than tomorrow's. Meanwhile, Gorbachev's generals may be given ever-increasing responsibilities to provide troops to help in civil reconstruction following earthquakes and in the building of new roads and towns.

Looking at his neighbors to the West, Gorbachev will undoubtedly breathe a sigh of relief as he leaves office. No longer will he be the Soviet leader who is walking a tightrope in trying to ensure a semblance of order within the Socialist bloc, where political and financial chaos increasingly threatens stability in most of the countries. With difficulty he should be able to superintend still another temporary truce with Solidarity, and he has little choice but to allow the Hungarians to stretch the limits of economic experimentation even further as their contribution to new thinking. The promised withdrawal of 50,000 Soviet soldiers and 5000 Soviet tanks from East Germany, Czechoslovakia, and Hungary presumably will take place; but the clamor from increasingly influential factions within Eastern Europe to remove all Soviet forces from the region may become loud and clear.

*   *   *

In Washington five years from now, Gorbachev will continue to witness government debates over how slow or how fast the United States should move toward a more relaxed and less adversarial relationship with the USSR. Many Americans will continue to believe that the changes in the USSR are the result of a one-man show, skeptical that a worthy successor to Gorbachev will be found and convinced that perestroika is reversible. The American military-industrial complex will be unhappy with any shifts in our military priorities away from the East-West confrontation; fighting terrorism and Third World insurgencies is not nearly as capital-intensive nor financially rewarding for entrenched economic interests as preparing for nuclear exchanges with the USSR.

Soviet-American relations will always be an important theme for many politicians and for the man on the street in the United States; but in the all-important arena of world economics, the Soviet Union will remain for the indefinite future a minor participant at best. Americans will continue to perceive the Soviet Union as a threat and NATO as an important force to repel the threat. But most Americans will become more concerned with the tightening encroachment of environmental decay around them and the next virus to follow AIDS into the mainstream of American life.

Gorbachev will soon see a Washington preoccupied as never before with population growth, poverty, and debt in the Third World; internal strife in Latin America and Asia; global environmental dangers; trade conflicts with Europe and Asia; and growing foreign ownership of American real estate. He will recall the Soviet mistake of selling Alaska to the United States, and he will warn us about our mistake of letting Japan gradually take over the Hawaiian Islands through real-estate acquisitions.

*   *   *

During the next five years, science and technology will be at the core of many policies adopted by the USSR to address their internal problems and to shape their foreign policies. Similarly, in the United States, science and technology will loom ever larger as we set the stage for entering the next century.

As we have seen, technology drives the military confrontation between the two countries while also providing powerful tools to monitor compliance with arms control agreements. Technology determines the rate and direction of economic growth in our two societies, although in the United States we are obviously on a more advanced plateau than are the Soviets. Unfortunately, many of the technologies that will determine our economic future are the same technologies that fuel the military confrontation; these dual uses are the root of our difficulties in devising collaborative efforts that will have economic payoff for both countries while helping to bridge political and cultural gaps.

One of the great challenges confronting the Bush administration is wise resolution of the cluster of technology issues concerning our trade and scientific relationships with the USSR, and particularly relationships involving dual-use technologies. While the National Security Council will have little difficulty developing broad policy directives advocating embargoes on technologies which are "critical" to maintain our military edge, translating such broad policies into hundreds of individual export control decisions will not be easy; the Pentagon considers many technologies critical, while American business considers very few to be so critical that we should forgo trade opportunities. Stricter enforcement of much shorter lists of embargoed items can better serve both interested parties. Advocating scientific cooperation in nonsensitive areas of mutual interest while restricting cooperation in sensitive areas also sounds good, but again the hundreds of decisions at the lower levels of the US Government on entry visas for Soviet scientists will be the important manifestation of our policies. If research programs are truly sensitive, they should be classified; visa restrictions will have little impact on preventing worldwide diffusion of unclassified research methods and results. Let us hope that our new officials will recognize that many past attempts to punish the Soviets by denying them access to our products and our laboratories have only punished the US economy and the advancement of science.

In one area, namely, the preservation of the global environment, we can help the Soviet Union and they can help us without becoming too deeply ensnarled in the problems of military technology. Many cooperative programs are underway bilaterally and through international organizations concerning the warming of the globe, preservation of important plants and animals that enrich the genetic pool of the planet, pollution of the oceans, and other ecological issues of broad international interest. Our scientific understanding of the causes and the extent of global environmental damage from human activity remains very rudimentary. Both countries have considerable scientific talent as well as much firsthand experience for addressing ecological global concerns, and the topic deserves a special status during discussions between the leaders of the United States and the USSR. Therefore, the two governments should amend their four-point political agenda (arms control, human rights, regional issues, bilateral cooperation) with a fifth topic, namely, global issues. While environmental concerns would be the centerpiece of this new agenda item, consideration of a broader range of energy issues and health problems which affect the entire world would very naturally enter the discussions and the calls for action.

Turning more directly to scientific cooperation, we have seen the benefits from many science ventures with the USSR, but we have also experienced the frustrations of wasted efforts. As Soviet scientists begin to put their huge house in better order, opportunities for shortcutting the road to scientific discovery through cooperative endeavors increase. We have identified many areas with high technical payoff through cooperation, while recognizing that there will be minimal technical benefit from cooperation in many other fields where the Soviets have little to offer.

However, even in some fields of current Soviet weakness, cooperation can help avoid surprises and can return other benefits as well. One such case is cooperation in the field of economics, where we have little to learn. Indeed, a few months ago the leading Soviet economist Leonid Abalkin told us that his colleagues have no idea what is meant by inflation; a Soviet colleague added, "We have one textbook on the topic, but it was written 50 years ago." Our staying abreast of Soviet thinking in this area, and indeed influencing Soviet thinking, is clearly in our national interest. Similarly in medicine, we may not want to adopt Soviet surgical procedures for reshaping the eyeball as an alternative to wearing glasses. But staying abreast of their assembly-line techniques for completing 15-minute, computer-assisted eye operations on hundreds of walk-in/walk-out patients every day may provide us with ideas for related approaches to medical problems afflicting tens of millions of Americans.

Effective scientific exchange programs require many years to develop and mature before significant benefits can be reaped. In the United States, only the US Government has the financial resources to support such long-term endeavors. Private foundations often provide short-term support for novel cooperative programs but seldom will consider support over many years. Therefore, the Bush administration should take steps to reverse the steady decline in government funds available for Soviet-American scientific cooperation—a decline that continued for eight years under the Reagan administration. By forgoing the purchase of one jet fighter every three years, we could double the current size of our scientific exchanges.

Finally, with regard to Eastern Europe, we can continue to benefit technically from cooperation with selected scientists of the region; and the political implications of such cooperation are increasing daily as the turmoil churns in the region. Still, US Government funding to support such cooperation is minimal. While our diplomats talk enthusiastically about the importance of cooperation with the region, American budget officials have yet to consider such cooperation as important as the purchase of two tanks each year.

*   *   *

Being a tennis buff, I carry my racquet to Moscow and the capitals of Eastern Europe. I sometimes play with tennis balls produced in the local factories, only to find that they usually bounce like sponges. When I complain to my partners that I can't reach the ball because it doesn't bounce very high, they simply reply that I must run faster. The good players run faster, and they win the points. But most of us simply curse the factory that produced such lousy balls.

In the USSR and in Eastern Europe the people have been expected to run faster and faster as the products of their factories became softer and softer. Now they are worn out from their running. The spirits of even the most avid players have been broken, and the facilities where they practice their trade are used less and less and gradually fall into disrepair.

Tennis players like to play well, but they also like to win and to take home the trophies. In the past, Soviets have worked hard in the laboratories, in the factories, and in the fields, but they have also wanted to receive recognition. Can Gorbachev help them receive this recognition so they will continue to try to win the games by running after the balls that don't bounce very high? Or should we provide them with balls that bounce a little higher so they can play at a higher level?

Yes, we should help Gorbachev and the Soviet people. We can help them in ways that advance our national security and that promote our human values. We not only have high-bouncing tennis balls, but we have many scientific and technological tools that can help them raise their standard of living.

Unlike the game of tennis, in international relations there need not be winners and losers. As American and Soviet scientists work together, the winners can far outnumber the losers as both countries gain. Many years ago Einstein recognized that "Nationalism is an infantile disease. It is the measles of mankind." Let us hope that our two countries are rapidly emerging from the adolescence of scientific nationalism.

Most importantly, American scientists have known and respected for many years Soviet scientific colleagues who are now members of Gorbachev's brain trust and who fully appreciate the need to redirect the Soviet scientific enterprise toward peaceful purposes. The likelihood of their success in doing so depends to a large extent on the US response to the new opportunities for relaxation of military tensions and for scientific cooperation. This is the challenge of techno-diplomacy.

Notes

PREFACE

1. Adapted from the definition of diplomacy set forth in *Webster's Ninth New Collegiate Dictionary*, Merriam-Webster, Inc., 1987, p. 357.
2. Letter from Franklin D. Roosevelt to Maxim M. Litvinov, Soviet People's Commissar for Foreign Affairs, November, 16, 1933.
3. The archives of US Government agencies, exchange organizations, and research centers bulge with eye-witness accounts of conditions in Soviet scientific research institutions. Productivity of Soviet scientists relative to Western counterparts is a popular topic in their reports. The estimate of 30 percent is based on private and public statements during late 1988 by officials of the Soviet Academy of Sciences and is generally consistent with such reports.
4. Secretary Shultz was particularly forceful on this point during a seminar at the National Academy of Sciences on September 14, 1987, devoted to the impact of information technologies on international relations.

CHAPTER 1

1. A particularly powerful film depicting the activities of Soviet gangsters is "Outside the Law," which was showing in Moscow in November 1988. The film concludes with the removal of a huge photograph of Brezhnev from a building in Moscow, apparently symbolizing the end of corruption within the police force, which had been ignoring criminal activities in exchange for

regular payments to senior police officials. Also during the fall of 1988, the Moscow weekly *Ogonyok* carried a number of articles about unpunished crime and corruption in the USSR.

2. See, for example, Patrick Cockburn, "RIP Kremlinology. With Glasnost Who Needs to Read Tea Leaves?" *The Washington Post*, February 14, 1988, p. C1.

3. Mikhail S. Gorbachev, *Perestroika, New Thinking for Our Country and the World*, Harper and Row, 1987. Two other publications also present his views, namely, Mikhail S. Gorbachev, *Mandate for Peace*, Paperjacks Ltd., New York, 1987, and Mikhail S. Gorbachev, *A Time for Peace*, Richardson and Steinman, New York, 1985.

4. Abel Aganbegyan, *The Economic Challenge of Perestroika*, Indiana University Press, 1988.

5. "Glasnost in the Soviet Union, An Update," *Soviet and East European Report*, Radio Free Europe/Radio Liberty, April 10, 1988.

6. Interesting commentaries on the development of the Soviet economy under Gorbachev, including management and international economic concerns, are included in Marshall Goldman, "Gorbachev, Turnaround CEO," *Harvard Business Review*, May-June 1988, pp. 107–113; Jerry F. Hough, "Opening Up the Soviet Economy," The Brookings Institution, 1988; "The Soviet Economy," *The Economist*, April 9, 1988, pp. 3–18; Ed A. Hewett, "The Foreign Economic Factor in Perestroika," *The Harriman Institute Forum*, August 1988; and Vladislav L. Malkevich, "The Importance of Trade in East-West Relations," *Journal of the US-USSR Trade and Economic Council*, vol. 13, no. 5, 1988.

7. In addition to the writings of Gorbachev, interesting perspectives on changes in Soviet strategies are included in Robert Levgold et al., "Gorbachev's Foreign Policy; How Should the United States Respond?" Foreign Policy Association, April 1988; and Rozanne Ridgway, "Perspectives on Change in the Soviet Union," Department of State, Current Policy No. 1090, July 1988.

8. See, for example, John Hardt, "Perestroika and Interdependence: Toward Modernization and Competitiveness," Presented in Seoul, Korea, on July 26, 1988, at a seminar sponsored by the George Washington University and the Korean Association of Communist Studies.

9. The rapid spread of technologies is discussed in *Globalization of Technology, International Perspectives*, Council of Academies of Engineering and Technological Societies, National Academy Press, 1988.

10. George Shultz, "National Success and International Stability in a Time of Change," Department of State, Current Policy No. 1029, December 1987.

11. George Shultz, "Managing the US-Soviet Relationship," Department of State, Current Policy No. 1129, November 1988.

12. A reasonably complete list of intergovernmental agreements in place in mid-1988 is set forth in *Newsletter*, The Soviet-East European Program of the National Academy of Sciences/National Research Council, Summer 1988, p. 17.

CHAPTER 2

1. Differing views of Soviet military doctrine, strategies, and activities are set forth in the following: Gareyev Makhmut, "The Revised Soviet Military Doctrine," *Bulletin of the Atomic Scientists*, December 1988; Dmitri Yazov, "The Soviet Proposal for European Security," *Bulletin of the Atomic Scientists*, September 1988, pp. 8–11; L. Ivanov, "Nuclear Weapons and Defense of Western Europe," *Pravda*, November 29, 1988, p. 6; Casper W. Weinberger, "Arms Reductions and Deterrence," *Foreign Affairs*, Spring 1988, pp. 700–719; Gerhard Wettig, "New Thinking in Security," *Problems of Communism*, March-April 1988, pp. 1–14; and William E. Odom, "Soviet Military Doctrine," *Foreign Affairs*, Winter 1988/1989, pp. 114–134.

2. "Western Defense, The European Role in NATO," EUROGROUP, Brussels, May 1988.

3. A recent US Government report on burden sharing is "Sharing the Roles, Risks, and Responsibilities for the Common Defense," A Report to the United States Congress, Department of Defense, December 22, 1988.

4. The Department of State summarizes the NATO approach to security in Ref. 2.

5. "Conventional Forces in Europe: The Facts," NATO, November 1988.

6. "Statement of the Warsaw Pact Defense Ministers Committee 'On the Correlation of Warsaw Pact and North Atlantic Alliance Force Strengths and Armaments in Europe and Adjoining Waters,'" *Pravda*, January 30, 1989, p. 5.

7. See Senator Carl Levin, "Beyond the Bean Count," Armed Forces Subcommittee on Conventional Forces and Alliance Defense, Second Edition, July 1988.

8. Mikhail S. Gorbachev, Speech to the United Nations, December 8, 1988, Soviet Mission to the United Nations. A presentation of the proposals is presented in "This Is Not Just a Matter of Tactics," *Newsweek*, December 19, 1988, pp. 31–32.

9. "Understanding the INF Treaty," US Arms Control and Disarmament Agency, 1988. Also, "The INF Treaty: Questions and Answers," Department of State, February 1988.

10. Many articles have been published during the past several years in *Bulletin of the Atomic Scientists* presenting all sides of the arguments concerning verification and the need for continued testing. The position of the Reagan administration on testing is set forth in "Nuclear Testing Limitations: US Policy and the Joint Verification Experiment," Department of State, July 1988; and "Reagan Administration Efforts in Nuclear Testing: A Chronology," *Arms Control Update*, US Arms Control and Disarmament Agency, August 1988. Also, a particularly succinct statement of the US position was

released by the US Delegation to the Nuclear Testing Talks in Geneva on September 17, 1987, namely:

> . . . a comprehensive ban on nuclear testing is a long-term objective, which must be viewed in the context of a time when we do not need to depend on nuclear deterrence to ensure international security and stability, and when we have achieved broad, deep and verifiable arms reductions, substantially improved verification capabilities, expanded confidence building measures, and greater balance in conventional forces.

11. Early commentaries on SDI and the Soviet response are included in "Soviet Strategic Defense Programs," Department of State and Department of Defense, October 1985; and R. Z. Sagdeyev and S. N. Rodionov, "Space Based Anti-Missile System: Capabilities Assessment," Space Research Institute, Academy of Sciences of the USSR, 1986. A more recent discussion is Benjamin Lambeth and Kevin Lewis, "The Kremlin and SDI," *Foreign Affairs*, Spring 1988, pp. 755–770.
12. An expansion of this theme can be found in Matthew Evangelista, "How Technology Fuels the Arms Race," *Technology Review*, July 1988, pp. 43–49.
13. For useful background, see *Reykjavik and Beyond, Deep Reductions in Strategic Nuclear Arsenals and the Future Direction of Arms Control*, Committee on International Security and Arms Control, National Academy of Sciences, National Academy Press, 1988.
14. Additional perspectives on the prospects for arms control agreements are set forth in George Shultz, "The Administration's Arms Control Agenda," Department of State, Current Policy No. 1121, November 1988; Jan M. Lodel, "An Arms Control Agenda," *Foreign Policy*, Fall 1988; and Leon V. Sigal and Jack Mendelsohn, "The Stage Shifts in Arms Control," *Technology Review*, August/September 1988, pp. 52–61.

CHAPTER 3

1. Roald Z. Sagdeyev, "Science and Perestroika: A Long Way To Go," *Issues in Science and Technology*, Summer 1988, pp. 48–52; R. Sagdeyev, "19th Party Conference: Tasks of Restructuring; Where Have We Lost Momentum?" *Izvestiya*, April 28, 1988, p. 3.
2. Among the many history books that describe this stage of development of the Soviet state is Nicholas V. Riasanovsky, *A History of Russia*, Fourth Edition, Oxford University Press, 1984.
3. Terrence Garrett, "Soviet Science and Technology—a New Era?" Department of Trade and Industry, London, February 1988, p. 1.
4. "From the Notes of Academician V. Legasov," *Pravda*, May 20, 1988, p. 3.
5. These and other Soviet building practices were discussed by American scientists and engineers who visited Armenia shortly after the earthquake

during a public meeting at the National Academy of Sciences on January 23, 1988.

6. Reliable information on Soviet expenditures for military research and development is not readily available. The estimates are based on a review of Soviet information on research expenditures in general and on information developed during Hearings of the Joint Economic Committee of the US Congress in the mid 1980s.

7. Garrett, p. 6.

8. "Decline of Prestige in the Engineering Profession," *Tekhnika i Nauka*, no. 11, November 1987, p. 10.

9. Louis Lavoie, "The Limits of Soviet Technology," *Technology Review*, November/December 1985, p. 70.

10. *Soviet Military Power: An Assessment of the Threat, 1988*, Department of Defense, April 1988, p. 141.

11. Some of these comments are based on a seminar held at the National Academy of Sciences on July 21, 1988, on the topic of dual-use technologies.

12. "The Soviet Economy," *The Economist*, April 9, 1988, p. 11.

13. Michael J. Berry, ed., *Science and Technology in the USSR*, Longman Group UK Ltd., 1988, p. 58.

14. Several recent publications provide interesting overviews of trends in Soviet science. They include: Loren R. Graham, "Gorbachev's Great Experiment," *Issues in Science and Technology*, Winter 1988, pp. 23–32; Craig Sinclair, "Reflections on Scientific Research in the Soviet Union," *Science and Public Policy*, June 1987, pp. 133–138; "Soviet Science," *Nature*, October 29, 1988, pp. 779–800; Richard M. Judy, "Soviet Science and Technology: Problems, Policies, and Directions," Prepared for the Subcommittee on Europe and the Middle East, House of Representatives Foreign Affairs Committee, US Congress, April 13, 1988; and "A Study of Soviet Science," US Government (released by the Office of Science and Technology Policy), December 1985.

CHAPTER 4

1. "Perestroika Throws a Party," *The Economist*, October 1988, p. 51.

2. Comments on reforms in the agricultural sector are included in Grant Mangold, "Portraits of Perestroika," *Soybean Digest*, December 1988, pp. 6–13; Karen Brooks, "Gorbachev Tries the Family Farm," *Bulletin of the Atomic Scientists*, December 1988; "Freeing Russia's Farms," *The Economist*, August 6, 1988, pp. 37–38; and M. A. Polyakov and A. N. Khitrov, "Economic Concepts of Scientific and Technological Progress in the USSR Agroindustrial Complex," 1988 (unpublished manuscript by two Soviet specialists).

3. Among the many discussions of Soviet economic reforms are "Gorbachev's Economic Plans," Joint Economic Committee, US Congress, vols. 1 and 2,

November 23, 1987; Ed A. Hewett, *Reforming the Soviet Economy*, The Brookings Institution, 1988; "The Soviet Economy," *The Economist*, April 9, 1988, pp. 3–18; Hedrick Smith, "On the Road with Gorbachev's Guru," *New York Times Magazine*, April 10, 1988; Richard E. Ericson, "The New Enterprise Law," *The Harriman Institute Forum*, Columbia University, February 1988; "Soviet Defense Ministry Boosts Output of Consumer Goods," *Izvestiya*, November 10, 1988; "United States-Soviet Relations: 1988," Hearings before the Subcommittee on Europe and the Middle East, House of Representatives Committee on Foreign Affairs, US Congress, February-April, 1988.

4. Insights concerning the cooperative movement are included in David Remnick, "For Kremlin, Coops Are Good—to a Point," *The Washington Post*, January 8, 1989, p. A23; and "The Growth of New Soviet Cooperatives," *Journal of the US-USSR Trade and Economic Council*, vol. 13, no. 5, 1988, pp. 35–43.

5. Reports on recent changes within the Soviet scientific establishment include Michael Parks, "Moscow Reprograms Its Scientists To Get Results," *Los Angeles Times*, October 18, 1988, p. 1; G. I. Marchuk, "Soviet Science Is the Breakthrough to the Future," *Ekonomicheskaya Gazeta*, no. 46, November 1987, pp. 20–21; G. I. Marchuk, Text of Speech at Party Conference, *Pravda*, July 1, 1988, p. 2; V. Konovalov, "The Scientists and the Time," Interview with G. I. Marchuk, *Izvestiya*, March 22, 1987, p. 2; V. A. Kirillin et al., "Elections to the Academy of Sciences," *Pravda*, December 18, 1987; Malcolm W. Browne, "Far Reaching Changes Seen in Soviet Science", *New York Times*, February 15, 1988, p. A22; and Roald Sagdeyev, "USSR Academy of Sciences at a Turning Point," *Moscow News*, January 3, 1988, p. 12.

6. Report of speech by G. I. Marchuk at June 29, 1988, session of 19th All-Union Communist Party Conference, *Pravda*, Second Edition, July 1, 1988, pp. 2–3.

CHAPTER 5

1. Additional comments on the relationship between exchanges and foreign policy can be found in Glenn E. Schweitzer, "Who Wins in US-Soviet Science Ventures?" *Bulletin of the Atomic Scientists*, October 1988, pp. 28–32; Yale Richmond, "Soviet-American Cultural Exchanges: Ripoff or Payoff?" The Wilson Center, Smithsonian Institution, November 1984; John M. Joyce, "US-Soviet Science Exchanges, A Foot in the Soviet Door," Russian Research Center, Harvard University, 1982; and "US-Soviet Exchanges: The Next Thirty Years," The Eisenhower World Affairs Institute, February 1, 1988.

CHAPTER 6

1. Many of the currently active joint ventures are identified in "Special Issue: Joint Ventures," *USSR Technology Update*, Delphic Associates, May 5, 1988.

Comments on future prospects are included in John E. Parsons, "The Future of East-West Industrial Cooperation," *Technology Review*, November/December 1988, pp. 57–63.

2. Prospects for cooperation in space research are discussed in Kathy Sawyer, "Taking Détente to Mars and Beyond," *The Washington Post*, May 15, 1988, p. B3; and Henry F. Cooper, "A Reporter at Large, Explorers," *The New Yorker*, March 7, 1988, pp. 43–61.

3. Two interesting reviews of scientific exchanges are "Review of US-USSR Interacademy Exchanges and Relations," National Research Council, September 1977; and Catherine P. Ailes and Arthur E. Pardee, "The US-USSR Agreement on Cooperation in the Fields of Science and Technology: Review and Evaluation," SRI International, March 1984.

4. Selected scientific benefits of exchanges are documented in "United States-Soviet Scientific Exchanges," Hearings of the Subcommittee on Europe and the Middle East, House of Representatives Committee on Foreign Affairs, US Congress, July 31, 1986; and "Selected Aspects of US-USSR Cooperation in Science," Prepared by the National Academy of Sciences/National Research Council for the Subcommittee on Europe and the Middle East, House of Representatives Committee on Foreign Affairs, US Congress, August 1986.

5. "Briefing Book for NAS Exchangees to the USSR," National Academy of Sciences, 1987/1988.

6. Another perspective on scientific relations between the two countries is presented in Kim McDonald, "US Researchers See Historic Shift in Relations with Soviet Scientists," *The Chronicle of Higher Education*, June 22, 1988, p. 1.

CHAPTER 7

1. "Soviet Acquisition of Militarily Significant Western Technology, an Update," US Government (department unspecified), September 1985, p. 2.

2. "To Examine US-Soviet Science and Technology Exchanges," Hearings before the Subcommittee on Science, Space, and Technology, US House of Representatives, June 23–25, 1987, p. 11.

3. Judith Axler Turner, "Soviet Espionage Efforts Have Targeted US Research Libraries and Staffs since 1962, FBI Charges in Report," *The Chronicle of Higher Education*, May 25, 1988, p. 1.

4. Brendan Greeley, "Soviets Target US Companies, Universities for New Technologies," *Aviation Week and Space Technology*, September 30, 1985, p. 86.

5. "Assessing the Effect of Technology Transfer on US/Western Security. A Defense Perspective," Office of the Undersecretary of Defense for Policy, February 1985, p. E-1.

6. Ibid., pp. 1–4.

7. Ibid., pp. 1–7.

8. Ibid. Among the many other sources is Michael Weisskopf, "Soviet Radar Allegedly Stolen from US," *The Washington Post*, September 24, 1985, p. A15.

9. See, for example, "Tracking a Techno-bandit," *Newsweek*, December 7, 1987, p. 66.

10. David E. Sanger, "In Shift, US Says Toshiba Sale to Moscow Was Damaging," *The Washington Post*, March 15, 1988, p. 8.

11. "Soviet Acquisition of Technology, an Update," p. 3.

12. Ibid, p. 4.

13. *Scientific Communication and National Security*, National Academy of Sciences, National Academy of Engineering, Institute of Medicine, National Academy Press, 1982.

14. The export control laws and regulations are detailed and complex. Of particular relevance to scientific cooperation is "Export Administration Regulations," Scientific or Educational Data, pt. 379, p. 2. Also, see "Export Control of Technical Data," International Trade Administration, Department of Commerce, July 20, 1983.

15. *Balancing the National Interest, US National Security, Export Controls, and Global Competition*, National Academy of Sciences, National Academy of Engineering, Institute of Medicine, National Academy Press, 1987; and *Global Trends in Computer Technology and Their Impact on Export Control*, National Research Council, National Academy Press, 1988.

16. *Scientific Communication and National Security*, p. 65.

17. For further insights on the complexities of technology transfer, see Bruce Parrott, ed., *Trade, Technology, and Soviet-American Relations*, Indiana University Press, 1985; Thane Gustafson, "Selling the Russians the Rope? Soviet Technology Policy and US Export Controls," Rand Corporation, April 1981; "East-West Technology Transfer: A Congressional Dialogue with the Reagan Administration,'" Joint Economic Committee, US Congress, December 19, 1984; Andrew C. Revkin, "Supercomputers and the Soviets," *Technology Review*, August/September 1986, p. 69; Janice R. Lang, "Scientific Freedom: Focus of National Security Controls Shifting," *Chemical and Engineering News*, July 1, 1985, p. 7; and Allen Wendt, "US Stance toward the Soviet Union on Trade and Technology," Department of State, Current Policy No. 1128, November 1988.

CHAPTER 8

1. Eduard Shevardnadze, Statement to UNESCO, USSR Mission to UNESCO, October 12, 1988.

2. "Reform and Human Rights: The Gorbachev Record," Commission on Security and Cooperation in Europe, US Congress, May 1988.

3. Robert B. Cullen, "Human Rights: A Millennial Year," *The Harriman Institute Forum*, Columbia University, December 1988.
4. For example, see Stephen White, John Gardner, and George Schopflin, "Democracy and Human Rights," *Communist Political Systems: An Introduction*, St. Martin's Press, 1982.
5. "Conference on Security and Cooperation in Europe: Final Act, Helsinki, 1975," Bureau of Public Affairs, Department of State.
6. See, for example, "Glasnost and State Secrets," *Soviet East European Report*, Radio Free Europe/Radio Liberty, April 10, 1988.
7. A discussion of Soviet psychiatric practices is included in Constance Holden, "Politics and Soviet Psychiatry," *Science*, February 5, 1988, p. 551. A more highly technical discussion is included in Walter Reich, "The World of Soviet Psychiatry," *The New York Times Magazine*, January 30, 1983, pp. 20–26.
8. See "The USSR Withdraws from the World Psychiatric Association," A Chronicle of Human Rights in the USSR, no. 48, October 1982–April 1983.
9. Current trends in modifying psychiatric practices are reported in Bill Keller, "Mental Patients To Get New Legal Rights," *New York Times*, January 5, 1988, p. A1.
10. Robert B. Cullen, "Soviet Jewry," *Foreign Affairs*, Winter 1986, pp. 252–266.
11. Additional discussions of problems in the USSR, including rumblings among the ethnic minorities, are reported in Douglas Stanglin and Jeff Trimble, "Raising the Stakes for a Good Czar," *US News and World Report*, November 28, 1988, p. 40; and "Cracks in the Soviet Facade," *Newsweek*, March 14, 1988, p. 26.

CHAPTER 9

1. *Eastern* Europe is a geopolitical definition of a region of countries of great diversity. The topography includes fertile agricultural valleys and barren hills, rugged mountains and coastal beaches, dense forests, large marshlands, rolling plains, and even an area of sand dunes. While most frequently considered as a political bloc, the countries have very different cultures, languages, and histories. Religious heritage and practice encompass large populations of Catholics, Moslems, and Orthodox religions and smaller populations of many other sects. Thus, the people are as different as the geography. Eight countries are considered to be in Eastern Europe based on their location and their embracement of communism since World War II. They are the so-called northern-tier countries of Poland, Czechoslovakia, and East Germany; the southern-tier countries of Hungary, Rumania, and Bulgaria; and two countries with unique political orientations, namely, Yugoslavia and

Albania. Historians have coined *Central* Europe as an area of political cohesiveness of the past, while military planners consider Central Europe to be a area of potential military conflict between East and West. The communist countries usually considered to be in Central Europe are Poland, Czechoslovakia, East Germany, and Hungary. Finally, in March 1988, Yugoslavia successfully convened a meeting of the leaders of the Balkan countries, reminding us that this geopolitical concept is still alive. These countries are Rumania, Bulgaria, Yugoslavia, Albania, Greece, and Turkey.

2. "Cracks in the Bloc," *Newsweek*, October 24, 1988, p. 30.
3. William Pfaff, "Reflections (Central and Eastern Europe)," *The New Yorker*, December 26, 1988, pp. 83–90.
4. "Cracks in the Bloc," p. 30.
5. A discussion of the organization and the strengths of science and technology in Eastern Europe is included in Gyorgy Darvas (ed.), *Science and Technology in Eastern Europe*, Longman Group UK Ltd., 1988.
6. For elaboration of developments in Bulgaria, see Glenn E. Schweitzer, "Introducing Research Results into Practice: The Bulgarian Experience," *Technology in Society*, vol. 9, 1987, pp. 141–155.
7. See, for example, "Eastern Europe Turns to West in Effort To End Technology Gap," *The Washington Post*, February 28, 1988, p. H1.
8. Discussions of US policy are included in John C. Whitehead, "The US Approach to Eastern Europe: A Fresh Look," Department of State, Current Policy No. 144, February 1988; and "US Policy toward Eastern Europe," Hearings before the Subcommittee on Europe and the Middle East, Committee on Foreign Affairs of the House of Representatives, US Congress, October 2–3, 1985.

CHAPTER 10

1. Robert G. Kaiser, "The USSR in Decline," *Foreign Affairs*, Winter 1988/1989, p. 97.
2. Comments on the outlook for perestroika are included in Leonid Abalkin, "Reconstruction in Management of the Soviet Economy, Preliminary Results and Prospects," Institute of Economics, Academy of Sciences of the USSR, February 22, 1988; Joel Kurtzman, "Of Perestroika, Prices, and Pessimism," *New York Times*, November 1988, p. 1; and "Economic Reform Stalled in USSR: Bureaucrats Manage to Keep Their Jobs," *Soviet East European Report*, Radio Free Europe/Radio Liberty, December 10, 1988.

Index